Stan Jacobs

The Adélie Penguin

The Adélie Penguin

Bellwether of Climate Change

David G. Ainley
with illustrations by Lucia deLeiris

COLUMBIA UNIVERSITY PRESS
NEW YORK

Columbia University Press
Publishers Since 1893
New York   Chichester, West Sussex
© 2002 David G. Ainley
All rights reserved

Library of Congress Cataloging-in-Publication Data
Ainley, David G.
   The Adélie penguin : bellwether of climate change / David G. Ainley.
     p. cm.
   Includes bibliographical references (p.).
   ISBN 0–231–12306–X (cloth : alk. paper)
   1. Adélie penguin—Climatic factors.   I. Title.

QL696.S473 A354 2002
598.47—dc21   2002024704

Columbia University Press books are printed on permanent and durable acid-free paper.

Printed in the United States of America

c 10 9 8 7 6 5 4 3 2 1

To Bill Sladen,
who had one foot on the heroic side of Antarctic exploration and the other in the modern scientific side. It was his passion and thirst for knowledge that helped lead us into the modern age of Antarctic ornithology.

Contents

Acknowledgments   xi
Outline of the Present Volume   xiii

Chapter 1. Introduction   1

Chapter 2. Marine Ecology   15
A Primer on Sea Ice As It Applies to Adélie Penguins   18
At-Sea Range and Habitat   25
Diet   43
Prey Capture   51
Foraging Behavior   55
Foraging Cycles and Energetics   65
How Much Prey Must Adélie Penguins Capture?   67

Chapter 3. Breeding Populations: Size and Distribution   69
What Is a Penguin Colony?   74
Adélie Penguin Colonies and Their Distribution in Antarctica   85
Geographic Structure   88
Factors Affecting the Distribution of Breeding Colonies   90

Chapter 4. The Annual Cycle   99

Colonial Nesting: Constraints and Advantages   *100*
Timing of Nesting: General Considerations   *102*
The Pattern of Colony Occupation   *106*
Timing of Nesting: Geographic Variation   *110*
Timing of Nesting: Effect of Age   *112*
The Molt   *121*
Further Thoughts About the Adélie Penguin's Annual Cycle   *127*

Chapter 5. The Occupation Period: Pair Formation, Egg Laying, and Incubation   131

Age at First Breeding   *131*
Nests and Territories   *136*
Pair Formation and Retention   *141*
Egg-Laying Date   *146*
Eggs   *150*
Incubation   *158*
Hatching Success   *165*
Further Thoughts   *170*

Chapter 6. The Reoccupation Period: Chicks and Breeding Success   171

The Chick   *172*
Phases of Parental Care   *175*
Chick Survival   *181*
Fledging   *183*
Breeding Success   *185*
What Breeding Patterns of Related Species Tell Us About Adélie Penguins   *191*
Reproductive Effort: Age Versus Experience   *196*

Chapter 7. Predation   199

Killer Whale   *200*
Leopard Seal   *203*
South Polar Skua   *210*

Chapter 8. Demography   219

Why Three-Year-Olds Risk Breeding   *222*
Fecundity   *224*

Survivorship and Age Structure   *226*
Why Is Breeding Deferred?   *232*
The Balance of Demographic Variables   *238*

CHAPTER 9. THE BELLWETHER OF CLIMATE CHANGE   243
Trends in Adélie Penguin Populations   *249*
Factors Explaining Population Trends   *253*
Patterns Revealed in the Prehistoric Record   *259*
Final Thoughts   *268*

Literature Cited   *271*
Index   *305*

Acknowledgments

A huge number of people have played a role in the publication of this book. The contributions and inspirations from Bill Sladen are immeasurable. Chuck Huntington introduced me to seabirds and their study, Robert Wood looked out for me at Cape Crozier, and Richard Penney showed me my first Adélie penguins and explained to me their displays. I spent many a day in the field and at school with Robert LeResche, comparing our notes and thoughts, hashing out the concepts that seemed to anchor how we perceive Adélie penguins. Doug DeMaster helped immensely in the quantitative analysis of the data. A number of friends have stood by me and inspired me in various important ways over the years in which this book was germinating: Grant Ballard, Jane Church, Rudi Ferris, Ian Gaffney, Robert Jones, Dick Mewaldt, Steve Morrell, Ed O'Connor, Susan Sanders, Larry Spear, Craig Strong, Michael Whitt, Peter Wilson, and, of course, my mom and dad.

I'm thankful to Lucia deLeiris for the fine artwork and to the editors of Columbia University Press—Alessandro Angelini, Holly Hodder, Jonathan Slutsky, Robin Smith, and Roy Thomas—as well as to my copyeditor Carol Anne Peschke, for their help in producing this volume. Several people read portions of the manuscript and offered many helpful comments: Lisa Ballance, Grant Ballard, Joanna Burger, Steve Emslie,

Stan Jacobs, Cynthia Tynan, and Rory Wilson. Others helped me to track down references: L. Ballance, A. Harding, G. Miller, M. Newcomer, R. Pitman, W. Sladen, and C. Tynan. The Raven Biological Library at Stanford University and the microfiche library at the Crary Laboratory, McMurdo Station, were tremendously helpful resources.

What could I have done without the folks from the National Science Foundation, Office of Polar Programs (NSF-OPP), and the U.S. Antarctic Program (USAP)? My Antarctic research has been funded by NSF-OPP, and the logistics were provided by USAP. The managers of the NSF Polar Biology Program helped out in many thoughtful ways: George Llano, Polly Penhale, Dick Williams, and Frank Williamson. The writing of this book was funded in part by NSF Grant OPP-9814882.

Outline of the Present Volume

This book is laid out as follows. Chapter 1 introduces Antarctica, the Adélie penguin, and early research on its natural history. Chapter 2 summarizes the marine ecology of this species, an area of research that has seen much progress in the past decade with the development and miniaturization of high-tech apparatus to study diving and foraging behavior. Chapter 3 summarizes the geographic structure of the Adélie penguin population: the size and distribution of its colonies and the factors that affect that distribution. Chapter 4 looks at the basic chronology of Adélie penguin nesting and population dynamics, as well as the molt. Chapters 5 and 6 bring in much material from an earlier monograph prepared by me and co-workers[1] but present it in the context of a large amount of additional information added by other researchers. Adélie penguins (and other penguin species) are unusual among seabirds in that they are not at the apex of the marine food chain, and chapter 7 describes the animals that prey on Adélie penguins. Chapter 8 is mostly a reprint, with some modification and addition, of the population biology and demography presented in the earlier volume.[1] Finally, chapter 9 provides a history of Adélie penguin populations both in recent decades and the period since the last glacial maximum 19,000 years ago. This last chapter offers hints as to how Adélie penguin populations may be changing in the future. The preceding chapters detail how those changes come about as this species copes with the Antarctic climate.

Throughout the volume, quantitative information is presented as simply as possible. All averages are given with the standard error (SE). This gives a measure of how different were the values used to calculate that average: A large SE value (e.g., equal to the average itself) means that the values were widely dispersed. Comparisons of averages often are made using a t statistic. This is merely a name for a type of statistical test (t values read from a table). When the t values are large, the difference in averages is statistically meaningful. Trends (changes in one value relative to another) are expressed using the value r, which is the regression coefficient. A high value of r means that the trend was statistically meaningful (e.g., the change in clutch size with greater age). A negative value of r means that one variable increases as the other decreases; a positive value means that both variables change in the same direction. Any statistic (t or r) is said to be meaningful if its probability, P, is less than .05. This means that there is a less than 1 in 20 chance that the value occurred only by happenstance, thus indicating a likely biological relationship. The only sophisticated statistical analyses are contained in the chapter on demography (ch. 8).

The pages of this volume are graced with the artwork of Lucia deLeiris. These sketches were published previously in a book by Sanford Moss and her, also produced by Columbia University Press.[13] Such artist-naturalists have carried on the traditions begun 100 years earlier by Edward Wilson, whose lovely paintings were published posthumously.

The Adélie Penguin

Chapter 1

Introduction

The spirit of adventure, under many guises, is what lures most people to Antarctica. Apart from tourism and industrial fishing, including whaling and sealing, commercial attractions are few and, in the case of mineral exploitation, only figments of the imagination at present. Crew members aboard ships might well have preferred to be any place but the Southern Ocean, yet the impetus behind such voyages stemmed from someone's appetite for adventure and discovery.

For me, Antarctica was a place to define myself against the life-defying forces of nature that are so evident there and known in the temperate zone only among the glaciers and peaks of temperate mountain ranges. As a teenager I had spent much of my time in the woods and fields near home. This soon led to hiking and climbing in the high mountains of western North America. Well before I became an undergraduate at Dickinson College, my interests had been leaning more toward the natural world. And once I discovered that one could make a living as a birdwatcher—after spending a summer as a National Science Foundation (NSF) undergraduate research assistant in the company of people studying seabirds on Kent Island, Bay of Fundy—I set my sights on Antarctica.

One of my heroes was Roald Amundsen (1872–1928), who from an early age also had set his sights on polar exploration. His accomplishments have always inspired in me a combination of reverence and excitement: the strong sense that whatever this explorer did, he could do it better than anyone. Amundsen led the first party to the South Pole. Ironically, for someone venturing into the unknown, his motto was "If you have had an adventure, then you were not prepared!" More than any of his peers, Amundsen appreciated the forces of nature he was dealing with.

Another of my heroes (and possibly a relative on my Grandmother Ross's side) is James Clark Ross (1800–1862), who in the 1840s led the first expedition to what would be called the Ross Sea, traveling significantly farther south than anyone before then and for the next sixty years. Many of the geographic features of the Ross Sea region were named during his two ventures there. A great deal of my own work in Antarctica has taken place on Ross Island, at Cape Crozier (named after Captain Francis R. M. Crozier of the *Terror,* the other ship in the Ross expedition; Ross was captain of the *Erebus*). The penguin colony at Cape Crozier lies in the shadow of Mt. Terror. Like Amundsen, Ross definitely knew what he was doing and prepared accordingly. He therefore encountered few "adventures"—and, unlike a number of polar explorers to come, never achieved the sensational fame so often surrounding misadventure.

In the introduction to his collection of writings by Antarctic explorers, Charles Neider expressed well what first drew me, and perhaps others, there:

> The lure of Antarctica, that fabulous, awesome and in some ways exquisitely beautiful last frontier, is so great that the long first leg . . . seems even longer despite the speed [of modern travel]. You travel with the suspense of a pilgrim eager to reach places sanctified by human and superhuman events. But the chief goal of the Antarctic pilgrim is to reach pure nature itself—from ancient, pre-human time, frozen in an incredible ice cap. Such a pilgrim goes to pay his respects to natural conditions; to take a breath of unpolluted air; and to sense how it all felt in the beginning, before the introduction of man. But he also goes to see how man survives in the world's most hostile place and does so by means of the very gadgetry which increasingly possesses and assails him. He travels, in short, from technological defilement for a glimpse of innocence, hoping to learn along the way a few things about himself. (pp. 4–5)[15]

This theme of life pitted against elemental hostility captured me, and after some twenty-two trips to Antarctica and its surrounding oceans, no one thing embodies the spirit of this struggle more than the Adélie penguin (*Pygoscelis adeliae*). Against incredible odds this little warm-blooded creature survives only by being supremely adapted, knowing what is important, and doing whatever is needed, always with amazingly defiant energy. This is especially true of this species' life in the ocean. There, huge waves toss blocks of ice, weighing many tons, this way and that. To capture prey, penguins must dive repeatedly for hours; just one dive lasts two to six minutes, often reaching depths of more than 150 meters. All of this is set against the specter of waiting leopard seals (*Hydrurga leptonyx*), who have no other thought than making a meal of the penguin.

I have always been amazed by the vivid change in demeanor exhibited each time an Adélie penguin comes ashore. This act takes incomparable athleticism. It's as if the penguin knows that all the cards are stacked against it, if not this time then certainly the next; if not a leopard seal, then huge waves; if not heaving blocks of ice, then an ice foot (the ice anchored to the beach, against which the pack or fast ice abuts; see chapter 7) necessitating a leap of two or more meters. The penguin lands ashore in a bad mood, exasperated and seemingly oblivious to what the beach has to offer. Then, as it shakes the water from its feathers and preens loft back into its coat, its awareness of the colony and of purpose seems to take over. It is only then that the penguin completes the transformation from marine to terrestrial creature. Off it goes, unquestioning and unvarying, sometimes even muttering almost inaudible renditions of calls it will use to greet mates or chicks.

llustration 1.1   Adélie penguins making their way across a snow field. *Drawing by Lucia deLeiris © 1988.*

Humans probably have known about Adélie penguins since the 1773–1775 voyage of James Cook, when on several occasions sea ice turned back Cook's ship, the *Endeavor*. On board was German naturalist George Forster, a great observer of nature, yet Forster made no specific note of this species.[20] Therefore it is difficult to pinpoint the first encounter. The introduction of Adélie penguins to humans may have occurred well before then, judging from what is known of Polynesian voyages, which reportedly reached southern sea ice during the mid-600s.[25] If any of these early voyagers did see penguins standing on ice floes in the southern reaches of the Pacific Ocean, they probably were Adélie penguins. On the other hand, these voyages were during summer, and we now know that Adélie penguins, particularly adults, are not plentiful along the outer edges of the pack ice during midsummer (or even, during winter, at the outermost edges); most dwell near to, if not at, their onshore colonies. Any Adélie penguin seen along the outer pack ice probably would have been a yearling, whose plumage is distinct from that of adults.

Not long after Cook's voyages, during the late 1700s and early 1800s, sealers visited Antarctic shores where today Adélie penguins are present (South Shetlands, South Orkneys). Explorers with less direct commercial interest also visited, such as Russian Commodore Thaddeus Bellingshausen, who in 1820 visited islands in the vicinity of the Antarctic Peninsula. Certainly these people would have encountered Adélie penguins and perhaps would even have had Adélie penguin served to them as food. Unfortunately, these early seal hunters or explorers made few, if any, notes on natural history.

THE WORD *penguin* may be derived from the Spanish *penguigo,* meaning *fat,* referring to the large quantity of fat found on auks, Northern Hemisphere counterparts to the penguin.[25] In Latin *pinguis* is the word meaning *fat.* The scientific name of the now-extinct great auk of the North Atlantic Ocean is *Pinguinus impennis.* Great auks and penguins, both flightless, are very similar in morphology, although not formally related (an example of what is called parallel or convergent evolution, that is, two unrelated organisms evolving in the same way to adapt to similar circumstances). Both the great auk and the first penguin known to science, the African penguin (*Spheniscus demersus*), were named by Carolus Linnaeus (although Linnaeus at first named the auk *Alca impennis*). Before Europeans spread themselves far afield, the indigenous people of southern Africa, Tierra del Fuego, New Zealand, and Australia no doubt knew of penguins. The Spanish and Portuguese probably were the first western Europeans to encounter penguins, which they did as they made their way on voyages around the Cape of Good Hope on their way to the East Indies in the 1500s. In southern Africa, these sailors would have encountered the African (also known as black-footed) penguin and soon afterward, when their routes took them west, the closely related Magellanic penguin (*Spheniscus magellanicus*) of Patagonia and Tierra del Fuego, including the Strait of Magellan.

It was not until January 20, 1840, that the Adélie penguin was described scientifically. At that time specimens were obtained, with proper descriptions then published in 1841 by Jacques Hombron and Honoré Jacquinot, surgeons doubling as naturalists on the French ships *Astrolabe* and *Zélée*. These ships were part of a French expedition led by Jules-Sébastien-César Dumont d'Urville. The expedition visited several Antarctic islands, including the South Orkneys and South Shetlands, and was the first to reach the Antarctic mainland—or at least some rocks offshore of what is now called Point Géologie, Adélie Land (66°33′S, 139°10′E). The latter section of Antarctic coast was named by Dumont d'Urville in honor of his wife, Adéle. The Adélie penguin was named after this place, where its breeding colonies were first visited by humans and critical scientific specimens were obtained.

For days the Dumont d'Urville expedition passed among ice floes teaming with Adélie penguins, but bird specimens were obtained only after the collection of rocks, proving that members of the expedition indeed had set foot on newly discovered land. The collection, with its disregard for the true citizens of that land, is described in the diary of Lieutenant Joseph-Fidèle-Eugène Dubouzet, executive officer of the *Zélée*.

> On these icebergs we saw flocks of penguins, which, unperturbed, stupidly, watched us pass by.
>   It was nearly 9 p.m. when, to our great delight, we landed on the western part of the highest and most westerly of the little islands. *Astrolabe*'s boat had arrived a moment before us; already the men from it had climbed up the steep sides of this rock. They hustled the penguins down, who were very surprised to find themselves so roughly dispossessed of the island of which they were the sole inhabitants. We immediately leapt ashore armed with picks and hammers. The surf made this operation very difficult. I was obliged to leave several men in the boat to keep it in place.

> I straight away sent one of sailors to plant the tricolour on this land that no human being before us had seen or set foot on.
>
> Following the ancient and lovingly preserved English custom, we took possession of it in the name of France, as well as of the adjacent coast where the ice had prevented a landing. Our enthusiasm and joy were boundless then because we felt we had just added a province to France by this peaceful conquest. If the abuses that have sometimes accompanied this act of taking possession of territory have often caused it to be derided as something worthless and faintly ridiculous, in this case we believed ourselves to have sufficient lawful right to keep up the ancient usage for our country. For we did not dispossess anyone. (cited in Rosenman, p. 474)[18]

Another description of this event—modern explorer meets (soon to be officially named) Adélie penguin—was given by Ensign Joseph-Antoine Duroch of the *Astrolabe*:

> There it is! We run ashore and to the sound of our joyful cheers, our brilliant colours unfurl and wave majestically below the Antarctic Circle above a boulder of rough reddish granite, overlooked by two hundred feet of eternal snow and ice.
>
> But we must have souvenirs; one of these fragments has to bring back to everyone of us in his old age the memory of stepping ashore on a new land. Picks and hammers ring out in competition. The rock is certainly hard, but it cannot resist our efforts, and soon many pieces fill the bottom of the boats.
>
> Some inoffensive penguins, the sole inhabitants of this place, are walking about near us and despite their protests we carry them off as living trophies of our discovery.
>
> But the wind is rising, as chill and cold as the ice over which it blows to reach us. We take advantage of it to get under sail and we salute the land as it disappears behind us with three cheers of "Vive le Roi!"
>
> The brisk wind drives us along smartly; at 11:30 p.m. we reach the corvettes where everyone is anxiously awaiting us on deck. At the sight of our trophies there is general rejoicing, our discovery is confirmed and receives the name Adélie Land. (cited in Rosenman, p. 478)[18]

In the area where these explorers landed, the Adélie penguin population is modest. On the rocks described, currently fewer than 1,000 pair breed, whereas in the general area of Point Géologie about 34,000 pairs are resident (see table 3.1). Almost a year later (January 12, 1841) and not far from where the French explorers came ashore, a party of British explorers, under Ross's leadership, also had a first encounter with nesting Adélie penguins. This time, however, the landing occurred within the species' population center, on an island offshore of what would become known as Victoria Land. Not surprisingly, the island's occupants left a greater impression on the intruders. Ross writes,

> The ceremony of taking possession of these newly discovered lands, in the name of our Most Gracious Sovereign, Queen Victoria, was immediately proceeded with: and on planting the flag of our country amidst the hearty cheers of our party, we drank to the health, long life, and happiness of Her Majesty and His Royal Highness Prince Albert. The island was named Possession Island. It is situated in lat. 71°56′, and long. 171°7′E., composed entirely of igneous rocks, and only accessible on its western side. We saw not the smallest appearance of vegetation, but inconceivable myriads of penguins completely and densely covered the whole surface of the island, along the ledges of the precipices, and even to the summits of the hills, attacking us vigorously as we waded through their ranks, and pecking at us with their sharp beaks, disputing possession; which, together with their loud coarse notes, and the unsupportable stench from the deep bed of guano, which had been forming for ages, and which may at some period be valuable to the agriculturists of our Australasian colonies, made us glad to get away again, after having loaded our boats with geological specimens and penguins. (1:189)[19]

At present, Possession Island hosts the world's second largest Adélie penguin colony—approximately 162,000 pairs—and lies within a day's sail of about a quarter of the world's Adélie penguin breeding population.

THE GENERIC NAME *Pygoscelis* is Greek for *rump-legged* (in reference to the upright, bipedal posture of a penguin) and was coined by German naturalist Johann Wagler, who in 1832 was considering the first-described member of this genus, the gentoo (*Pygoscelis papua*). Bones of this bird arrived in Europe among artifacts from Papua New Guinea (hence the anomalous derivation of the species name). The Adélie penguin was not considered a member of the *Pygoscelis* group until 1898, when William Ogilvie-Grant of the British Museum of Natural History argued successfully for its inclusion. Until then, it had been known scientifically as *Catarrhactes adeliae,* as originally described by Hombron and Jacquinot in 1841. Because the Adélie beak (fully feathered almost to the tip; no visible nostrils) and other aspects of plumage differed from those of the gentoo, in 1937 Robert Falla[5] argued unsuccessfully that the species be separated from *Pygoscelis* and known as *Dasyramphus adeliae.* Just a few years later, Brian Roberts[17] proposed that the Adélie penguin be placed in the genus *Pucheramphus,* a name derived from the naturalist who coined *Dasyramphus,* Jacques Pucheran. William Sladen[23] initially followed suit, but eventually he conceded to the wisdom of practicing taxonomists (e.g., Robert Murphy) and reverted thereafter to *Pygoscelis.*[24] The nomenclature referring to this bird has since become firmly accepted. The Adélie penguin has also been known commonly as the black-throated penguin, long-tailed penguin, or Adelia penguin.[14]

As noted by John Sparks and Tony Soper in their review of the penguin family of birds, the Adélie penguin is "the most familiar of all penguins, the smart little man in evening dress" (p. 212).[25] Murphy, probably the source of Sparks and Soper's view, elaborated its status among penguins in greater detail:

> The Adélie is the most thoroughly investigated of penguins. Because of the widespread publicity its arresting appearance and manners have been given by means of pen, camera, cinematograph, and even the radio, it has become, moreover, easily the most familiar of penguins. Popularly speaking, it is the type and epitome of the penguin family, a prestige developed entirely during the period of active south-polar exploration that began after the opening of the present [twentieth] century.
>
> With singular unanimity, explorers have likened the Adélie penguin to a smart and fussy little man in evening clothes, with the tail of the black coat dragging on the ground, and who walks with the roll and swagger of an old salt just ashore from a long voyage. (p. 387)[14]

Certainly, then, the Adélie is the best known of the pygoscelid penguins and among the best known of all penguins because of the several British expeditions that visited this species' stronghold in the Ross Sea and the naturalists who participated.[9,10,19,21,30] Complementary observations of this species were also made by early French,[6] Scottish,[4] and Swedish expeditions[2] to the Antarctic Peninsula.

Early Antarctic naturalists firmly established the overall patterns of

Adélie penguin breeding biology—the timing of breeding, the numbers of eggs laid, and the proportions of chicks hatched and fledged—and the penguins' dealings with skuas and leopard seals. Much of the writing dwelt on the "home life" of the penguins, with much description of mating ("love habits") and the contests for territories and mates. These early writers had a good deal to say about how humans and penguins responded to one another. Their writing was very colorful. James Murray, a biologist on Sir Ernest Shackleton's first expedition, describes the essence of Adélie penguins in typical fashion of the times as he tries to convey the Antarctic experience to a learned audience:

> The Adélie is very brave in the breeding-season. His is true courage, not the courage of ignorance, for after he has learned to know man, and fear him, he remains to defend the nest against any odds. When walking among the nests one is assailed on all sides by powerful bills. Most of the birds sit still on the nests, but the more pugnacious ones run at you from a distance and often take you unawares. We wore for protection long felt boots reaching well above the knee. Some of the clever ones knew that they were wasting their efforts on the felt boots, and would come up behind, hop up and seize the skin above the boot, and hang on tight, beating with their wings. (cited in Shackleton, p. 349)[21]

These early naturalists also observed that the Adélie penguin wintered on the outer edge of the pack ice because the light was sufficient for them to see;[5,14] that they bred in windswept areas to avoid snowdrifts and have access to pebbles and on hummocks to avoid meltwater;[9] that the adults returned year after year to their colonies and former nest sites; and that once fledged they did not reappear at colonies for at least two years.[6,9] Although they killed and ate many penguins and thus had ample opportunity for dissection, most early naturalists did not seem to notice that it is the male who arrives first and establishes the territory, does the bulk of nest building, and takes the first watch during incubation. The Swedish naturalists did notice this,[2] but others thought all these activities were performed by the female.[9,14,21]

Famed polar naturalist L. Harrison Matthews had an interesting if cynical opinion of the early work on Adélie penguin biology:

> The Adélie is the [penguin] best known to the general public because it is plentiful in the Ross Sea area where the expeditions of the early years of

this [twentieth] century worked, particularly those of Scott and Shackleton. The birds and their behaviour are so eye-catching, and they are so easily seen by uncritical observers as funny little manikins, that an enormous mass of anthropomorphic nonsense has been written about them. It is sad to think that for the great half-literate public, all the labours, all the hardships, all the successes and failures of the men who lived and died on those expeditions, have produced nothing beyond the popular "image" of the penguin, a ridiculous distorted travesty, and a soft toy for babies to cuddle in bed. It is that golliwog eye that does it. (p. 98)[11]

A great deal of fanfare has been bestowed on polar expeditions and the heroic feats of the participants, as nature adventures are published to satisfy an urban public thirsting for natural experience. In the process, the public perception of the early expeditions has moved closer to Matthews' views.

Matthews[11] correctly pointed out that it was systematic observations by a British medical officer and naturalist, William Sladen[24]—who made his first observations in 1948, overwintering at Hope Bay in 1948–49 and at Signy Island in 1950–51—that settled many differences of opinion, elucidated the basic breeding biology of this species, and supplanted the "anthropomorphic nonsense." Actually, leaving England in December 1947, he was supposed to study emperor penguins at Dion Island, but the ship, *John Biscoe,* could not get through the consolidated pack ice to reach shore (and therefore could not relieve the party of scientists there). Sladen accomplished much by banding penguins and closely following them through consecutive breeding seasons. He also took detailed notes describing the sex and physiological maturity of the penguins his party killed to feed their sled dogs. The latter was necessary when a fire destroyed the expeditions' hut and most of its contents a month after the ship left the group, not to return for two more months. Sladen distinguished the males from the females and roughly categorized age classes based on the maturity of behavior and physiology. He also figured out that parents fed only their own chicks. His treatise on Adélie penguin biology is a classic, as are his films documenting this species' natural history. In an innovative move, he submitted a film, *Life History of the Adélie Penguin,* with his thesis for his doctorate at Oxford University. Sladen's groundbreaking work introduced modern ornithology to the Antarctic, leading the transition from the heroic age of exploration.

Other ornithological pioneers of that era include Bernard Stonehouse, a meteorologist in the Falkland Islands Dependencies Surveys (as the British Antarctic effort was called at the time) and Jean Sapin-Jaloustre, an ornithologist with the first French Antarctic Research Expedition. Stonehouse contributed some of the first data on bird species that nest at South Georgia, Marguerite Bay, and the tip of the Antarctic Peninsula, such as the emperor penguin (*Aptenodytes forsteri*),[26] king penguin (*Aptenodytes patagonica*),[28] and brown skua (*Catharacta linnbergi*).[27] As discussed later in this volume, his later work included important contributions to our knowledge about Adélie penguins. Sapin-Jaloustre,[20] contributing much information about the social behavior of Adélie penguins, worked at what is now a major French research station, Dumont d'Urville, in Adélie Land, 1948–51.

In 1957–58, the International Geophysical Year (IGY), multinational research programs attempted to investigate many of the leading atmospheric, geologic, and geographic questions of that day. Several nations launched major expeditions representing various scientific disciplines to the Antarctic. A noteworthy contribution to our understanding of the Adélie penguin emerged from the IGY. Working at Cape Royds, Ross Island, Rowland Taylor[29] described in great detail the breeding ecology of the species, taking up where Sladen left off. Taylor also used banded birds and marked nests to aid his research. He may have been among the last people to occupy Shackleton's hut for the purpose Shackleton[21] intended: "to provide harbor from the blizzards for those who come after" (p. 378). Today, Shackleton's hut is locked to protect it from both tourists and scientists. Also during the IGY, Sladen visited sites in the Ross Sea hoping to set up a long-term study of Adélie penguins, which he had failed to accomplish at Signy Island.

Since the IGY, scientists from the United States, New Zealand, and a few other countries have studied Adélie penguins at colonies on Ross Island almost every summer. Investigations of natural history, demography, social displays, and physiology have been completed, in some cases more than once or from differing perspectives. Similarly, much has been learned about the congeneric gentoo penguin, beginning with studies in the very early days of Antarctic exploration[3,6,12] and culminating in a number of studies during the past twenty years as more countries have established bases on the Antarctic Peninsula. These bases are located in the exposed, gently sloped areas where gentoo and other penguins have established colonies. However, very little effort has been directed toward investigating (or at least describing) the natural history of the chinstrap penguin (*Py-*

goscelis antarctica), probably because this species nests primarily on steep, isolated islands of the Scotia Arc and the outer, more hazardous coasts of the South Shetland Islands and Antarctic Peninsula. This lack of interest is unfortunate because the chinstrap appears to be much more similar to the Adélie penguin than is the gentoo, and a comparison of how each species has solved the environmental problems it faces offers clues about the adaptations by the other. Ironically, William Sladen has a huge amount of data on chinstrap penguins at the South Orkneys but has never had the time to write about that species. To some degree, the chapters in this text exploit what is known about gentoo or chinstrap penguins to better describe the Adélie penguin.

Finally, the landmark Convention for the Conservation of Antarctic Marine Living Resources (CCAMLR), signed in the early 1980s,[8] has brought a number of other researchers and institutions to investigate the Adélie penguin. Increased penguin research has been justified by designation of the Adélie penguin as an indicator species. Changes in certain aspects of this species' breeding biology (e.g., breeding success, chick growth rates, and diet) are believed to be sensitive indicators of ecosystem change. Long-term data sets describing annual variation in this species' natural history are being accumulated by Australian, French, and Japanese research programs in East Antarctica, New Zealand and Italian programs in the Ross Sea, and British and U.S. programs in the Antarctic Peninsula.

This volume is intended to accomplish two goals. The lesser one is to present portions of the out-of-print volume written by me, Robert LeResche, and William Sladen,[1] *The Breeding Biology of the Adélie Penguin.* Many people have asked me where they can obtain a copy of that volume, but it is no longer available except, rarely, in used bookstores. LeResche and I were graduate students of Sladen, and this work combined our theses, which stemmed from the long-term banding study initiated by Sladen and Robert Wood at Cape Crozier in 1961. That banding went on for ten years, with 5,000 chicks given individually numbered flipper bands each year. LeResche and I came along at the end of that banding program, reaping the scientific benefits of all those known-age, known-historied birds. No comparable demographic information has since been gathered about this species. The long-term banding effort was Bill Sladen's dream, inspired by his earlier efforts in the Antarctic and by the long-term (eighteen-year) banding study conducted by Lance Richdale,[16] *A Population Study of Penguins,* on the sedentary (as opposed to the migratory

Adélie) yellow-eyed penguin (*Megadyptes antipodes*). This penguin species also was first described by French naturalists and surgeons J. Hombron and H. Jacquinot. Sladen and Richdale were contemporaries at Oxford University. They analyzed and wrote up their respective data sets, acquired in the same era in different corners of the Southern Hemisphere.

The greater goal of the present work is to summarize the huge amount of information that has been written about the Adélie penguin. Material from *The Breeding Biology of the Adélie Penguin* is presented here in the context of many other studies. Little about the Adélie penguin's social behavior is contained in the present volume. Through interspecific and intraspecific comparisons, new aspects of this species' natural history are revealed.

Indeed, the Adélie penguin is one of the best-studied avian species. As noted earlier, the investigation of its biology has been intertwined with the heroic exploratory expeditions to the Antarctic, the establishment of bases and research programs, and the recent establishment of monitoring programs under CCAMLR. This long history of research is fortunate because changes in this species' natural history patterns are now revealing how recently accelerated climatic change is affecting one of the largest surficial habitats on Earth, the sea-ice zone that rings Antarctica (40 percent of the Southern Ocean and 6 percent of the world's oceans).[7]

The Adélie penguin will keep fighting for survival as long as sea ice persists. Unfortunately, their world is being affected negatively by the climate changes we have contributed to and are experiencing at midlatitudes. Global warming is affecting polar regions much more acutely than lower latitudes; most readily affected by warming is the extent of sea ice, and this habitat is receding. In turn, the loss of sea ice accelerates warming because a significant amount of sunlight is no longer reflected away and therefore is absorbed by the polar oceans.

Lord Keith Shackleton recently wrote,

> In the early days of exploration Antarctica lay far from the centre of world affairs. I sometimes wonder if, during the first decade of the [twentieth] century, my father or any of the pioneer explorers imagined that, before the end of the century, their chosen region would attain such far-reaching global significance—as the scene of intense, large-scale scientific research, and the subject of a remarkable international treaty for peaceful research. . . . Perhaps they did: as visionaries who looked beyond the horizon. . . .

> It is now generally realised that we have the capacity to change our environment on a truly global scale. Environmental issues such as ozone depletion, enhancement of the atmosphere's greenhouse effect by industrial activities, and possible changes in sea level have assumed prominence on the world stage. Rightly so: the implications of these changes are far-reaching for almost all forms of life on this planet. If we are to anticipate and respond to the effects of human-induced climatic changes, it is vital that we understand the processes involved. (pp. ix-x)[22]

That huge heat sink, Antarctica, may resist major climate change the longest, but as the following chapters detail, the Adélie penguin is showing us that no habitat on Earth will escape. As noted earlier (and described in detail in ch. 9), the penguins' sea-ice habitat, which surrounds Antarctica, is showing the effects of global warming. My intent is to demonstrate that the penguin can help us to understand climate-changing processes. The chapters that follow explore the many ways in which sea ice, that unique attribute of polar seas, affects this species' well-being.

Chapter 2

Marine Ecology

Penguins are the most specialized and most capable divers among birds. Much of the foraging behavior of penguins has been studied, especially by biologists using instruments and apparatus developed to quantify aspects of diving, swimming, prey capture, and energetics. More than specialized divers, however, penguins truly are marine creatures, perhaps more so than any other group of birds. Therefore, to understand the Adélie penguin's behaviors at sea, where it spends more than 90 percent of its total life,[1] more than just a description of prey searching and capture is needed.

To understand the marine ecology of a penguin one must consider a number of topics, including at-sea geographic range, habitat preferences, offshore foraging areas and foraging range of breeders, proximity of nonbreeders to colonies during the breeding season, juxtaposition of breeding sites to areas inhabited during the annual molt, distribution during the winter and diet at all times of the year. There has been no effort to compile a total picture of these aspects of the Adélie penguin's marine ecology in any one region. Indeed, the information available is piecemeal at best, one aspect well researched here and another aspect there. Some aspects have been researched a great deal, others hardly at all. Here I attempt to synthesize the available information and make some inferences on obscure aspects of this species' attributes as a marine creature.

Actually, the marine ecology of the Adélie penguin is better known than it is for most penguin species, or indeed any other seabird species, but much more is known in this regard about the African penguin. The marine ecology of the latter was investigated intensively as part of the Benguela Current Ecosystem Project of the South African Fisheries Agency during the 1980s. This work included both at-sea and colony-based studies, and, perhaps fortuitously, Rory Wilson became involved in that effort. Wilson is a pioneer without equal in developing instruments and interpreting the data gathered on the foraging and diving behavior of penguins and other seabirds. Fortunately, he has conducted some work on the foraging behavior of the Adélie penguin. Jerry Kooyman, Boris Culik, Mark Chappell, and associates also have worked intensively to develop new insights into the diving physiology and foraging energetics of penguins, and this information can be used to further our appreciation of the Adélie penguin's life at sea.

Penguins are the quintessential marine birds. The Adélie penguin spends 90 percent or more of its life in or on the sea, be it liquid or frozen.[5] Other species, such as some of the crested penguins (*Eudyptes*), without ice floes on which to perch from time to time, are at sea so long during their annual cycle that barnacles have been found on their feet.

The key to understanding the marine ecology of the Adélie penguin is to realize that this species is a creature of the Antarctic pack ice. In that respect it is unlike any other penguin species except the emperor penguin. An analysis of the spatial and temporal persistence of the associations between seabird species in the South Pacific and adjacent Southern Ocean (conducted on the basis of cruises between the equator and the shores of Antarctica) found that the most invariant assemblage, and one having almost no overlap with others, is that of the Antarctic pack-ice zone.[80] The major species in that pack-ice assemblage are the Adélie and emperor penguins, the snow petrel (*Pagodroma nivea*) and Antarctic petrel (*Thalassoica antarctica*), and, in summer, the South Polar skua (*Catharacta maccormicki*). A few other species join this assemblage during summer, but the first four rarely stray far from the sea ice or drift ice (icebergs and the ice rubble resulting from their disintegration) at any time of year.

This bit of polar natural history was significant to the sealers, whalers, and adventurers who made the first excursions into the Southern Ocean a century or more ago.[66] The appearance of the "ice birds" meant that pack ice was not far away,[10,83] and any further southward progress was unlikely (figs. 2.1 and 2.2). The pack ice that rings the Antarctic conti-

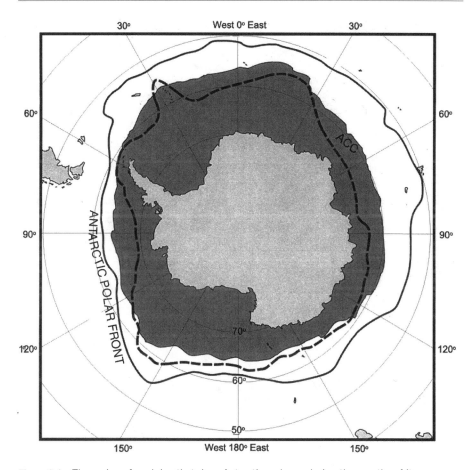

Figure 2.1 The region of pack ice that rings Antarctica, shown during the months of its maximal extent, usually July-September.[118] *Also shown:* Antarctic Polar Front[64] and southern boundary of the Antarctic Circumpolar Current (ACC).[70]

nent kept the wooden sailing ships of that day away from Antarctica.[40] To this day, most ship travel in the high-latitude Southern Ocean (almost always on strong-hulled ships or ice breakers) takes place where and when the sea ice is minimal. For various possible reasons we will consider in this chapter, the outer edge of the Southern Ocean pack ice also impedes the northward travels of the Adélie penguin.

The pack ice of the Southern Ocean is one of Earth's largest habitats. At its maximal extent, usually reached between July and September, the

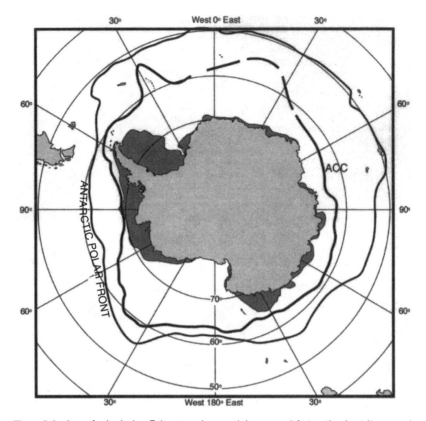

Figure 2.2 Ice refugia during February, when pack ice around Antarctica is at its annual minimal extent.[118] *Also shown:* Antarctic Polar Front[64] and southern boundary of the Antarctic Circumpolar Current (ACC).[70]

Antarctic pack ice covers 6 percent of the world's oceans (40 percent of the Southern Ocean[39]), with ocean covering three-fourths of the planet. A basic description of sea ice is needed before we continue our exploration of the Adélie penguin's marine ecology.

■ A Primer on Sea Ice As It Applies to Adélie Penguins

When the water at the sea surface reaches −1.8°C, it begins to freeze. This is a lower temperature than is needed to freeze the waters of lakes and streams because the salt dissolved in seawater disrupts the formation of ice crystals. Initially the water forms a slurry, but with further cooling some of the sea salt is squeezed out as the ice crystals develop further. The brine drains from the forming ice, making the underlying ocean more saline (scientists call this process salt rejection), and the slurry (frazil ice) begins to congeal into slush (nilas). With lower temperatures and further salt rejection, the nilas becomes solid enough for penguins, then larger animals, to walk on. This ice, called new ice, is 5–15 centimeters thick and is somewhat transparent and blue or gray (the color of the water below or the sky above, depending on the reflection of light; see fig. 2.3).

The new-ice stage is perilous for Adélie penguins: although it is strong enough to support the weight of one or more penguins, predatory leopard seals can see them from below. Moreover, the seals can break through to grab the penguins. James Clark Ross, in command of the sailing ship *Erebus,* describes these properties of new ice:

> As soon as we had completed all the necessary observations at this interesting spot [vicinity Coulman Island, Victoria Land], we commenced the laborious work of retracing our way through the pack to the eastward; but the young ice had so greatly increased in thickness, that this was a measure of great difficulty, and for a long time we had great doubts whether it would not prove too strong for us, and that in spite of our utmost exertions we might be frozen fast; for when we got clear of the heavy pack, the whole surface of the sea presented to our view one continuous sheet of ice, through which, when the breeze freshened up, we made some way, but were sometimes more than an hour getting a few yards; the boats were lowered down, and hauled out upon each bow, and breaking up the young ice by rolling them, we found the most effectual means: for although it was sufficiently strong to prevent our ships sailing through it, yet it was not strong enough to bear the weight of a party of men to cut a passage with saws. (I: 248)[83]

The plasticity of new ice is also important to Adélie (and emperor) penguins. Even where thickest, it is soft enough that minke whales (*Balaenoptera bonarensis*) can break through to breathe. This provides breathing holes between which penguins and seals can swim, thereby allowing them to escape rapidly from the early winter sea-ice freeze-up.

Where waves and swells are present, slushy new ice breaks into pieces that take on a disklike form because of the constant buffeting. From constant collision, the edges of each disk are turned up to form a lip, becom-

ing pancake ice (fig. 2.3). The pancakes grow in breadth as the freezing process continues, increasing from a few centimeters to a meter or two across. If it snows, the snowmelt collects in the rims, forcing salt downward. Finally, usually with a slackening of wave action, the pancakes stick together to form a rough-surfaced sheet. This conglomeration of ice disks is still plastic enough to undulate as the ocean swells travel beneath. Nevertheless, the speed at which this process can occur, especially with the rapidly decreasing temperatures of autumn, can have grave consequences for air-breathing creatures.

The dramatic suddenness of the freeze-up is evident in the diary of Captain Ross, one of the first people to witness this event in the Southern Ocean (or at least the first to write about it). Close to the date (mid-February) and location (77°56′S, 179°45′W) where I experienced the freeze-up—in the southeastern Ross Sea near the Ross Ice Shelf—some 150 years earlier, he wrote,

> Young ice formed so quickly in this sheltered position, and the whole space between the barrier [Ross Ice Shelf] and the main pack[,] which was driving down upon us[,] being occupied by pancake ice, we found ourselves in a situation of much difficulty. . . . At noon we were . . . very much hampered by the newly formed ice, which was so thick, and extended so far from the main pack, as to render our efforts to examine it quite fruitless, and the fatigue and labour excessive. We continued, however, to coast along its western edge seeking for an opening; but the severe cold of the last few days had completely cemented it together, and the thick covering of snow that had fallen had united it, to appearance, into a solid unbroken mass: although we knew quite well that it consisted entirely of loose pieces, through which only a few days before we had sailed upwards of fifty miles, yet we could find no part of it now in which we could have forced the ships their own length. (I: 237, 239)[83]

More salt is rejected from the sea ice with further freezing (and with melting of snow on the upper surface). The ice becomes opaque, and its

Figure 2.3 (opposite) Sea ice in various stages as it affects Adélie penguins. *Upper right:* frazil ice with steam from the heat released as the sea freezes; *upper left:* a large flock of penguins making their way across new ice (dark patch); *mid-right:* a flock of penguins resting on a continuous sheet of sea ice stretching to the horizon; *mid-left:* pancake ice; *lower right:* penguins making their way across a pressure ridge; *lower left:* penguins jumping from floe to floe.

plasticity all but disappears. It then reaches a point, about a half-meter thick, when any ocean swells cause a sheet of sea ice to fracture into ice floes (fig. 2.3). In Ross's words, "In those regions we have also witnessed the almost magical power of the sea in breaking up land ice or extensive floes, . . . which have in a few minutes after the swell reached them, been broken up into small fragments by the power of the waves" (I: 228).[83] About 10–20 kilometers into the ice pack, depending on the proportion of the ocean surface covered by ice floes, the weight of the ice dampens the motion of ocean swells, and with further distance from the open sea they are extinguished. Thus, deeper into the ice, small pancakes and floes become less common, and sheets are the rule. A shearing pressure from the wind blowing over a huge expanse of ice or blowing a sheet of sea ice against an iceberg (on which the wind has little influence) also can cause the ice to fracture into floes. As we shall see, a sea covered by ice floes with stretches of open water between them (called leads) is the ideal habitat of the Adélie penguin in spring, summer, winter, or fall.

Ice floes can range in expanse from a few meters to the size of large towns. The smallest pieces are called ice cakes or brash; the largest are called sheets. In the interior of the ice pack, where the effects of waves and wind are minimal or nonexistent, the Antarctic sea ice becomes a few meters thick over the course of the winter. Unlike most lakes, which could freeze solid to the bottom if temperatures were cold enough for a long enough time, the ocean is deep enough that with the greater pressure of depth and the buoyancy of ice itself, any further thickening of sea ice forces it to float higher in the water. As it rises, salt continues to drain from the upper layers, so that the upper surface of sea ice becomes almost fresh and is rock hard (hard enough to land our biggest airplanes on). At the undersurface there is enough salt or the freezing stage is early enough that the ice is still slushy. The rejected salt forms little channels, called brine channels.

The size of ice floes and sheets affects the activities of Adélie penguins, and the slushy underside, with its channels, provides habitat for some of the Adélie penguin's prey. Where the sea surface is covered by ice floes (or even brash), with up to 90 percent cover (i.e., there are still open-water leads between floes), it is called pack ice (fig. 2.3). The name comes from the fact that the floes can be packed together by the wind. If the wind is strong enough—and this is common in the Southern Ocean—the floes are forced over or under neighboring floes, creating consolidated pack, with few leads. In this way, ships traveling through leads can be crushed as the wind packs the floes together.[40] The juxtaposed floes then freeze together, forming a jumble of ice ridges above and below the ice surface. These pressure ridges form where ice floes collide; where floes are forced almost vertically, these ridges may be several meters high. Consolidated pack is rough and slow going for Adélie penguins, who have to climb up and down the ridges. Their long toenails aid in climbing, but their legs are only several centimeters long.

Where there are bays or sounds along the coast of Antarctica, the sea ice freezes into an uninterrupted sheet kept fast by islands, grounded icebergs, or points of land (fig. 2.4). This type of sea ice is called fast ice (held fast to the land). Its only cracks, caused by the rise and fall of the tide, are those along the shore and around icebergs (and islands). Fast ice can remain for most of the winter and spring, it can be permanent, or, more commonly in Antarctica, it can remain for a few years at a time. Then, a warm summer softens it sufficiently that ocean swells break it into floes and winds carry it away. Fast ice is a boon to humans in Antarctica: we can drive vehicles or land airplanes on it or we can drill holes through it to lower instruments, fish traps, or ourselves to collect data. As a result, we know a great deal about the ecology of the ocean beneath fast ice in the Antarctic, unlike the pack ice-covered portions of the Southern Ocean.[87] Fast ice is an impediment to Adélie penguins because to negotiate it they must walk, a mode of transportation much less energy efficient than swimming.[28,107] Where there is extensive fast ice (more than 10 kilometers or so wide between the shore and pack ice-covered seas) remaining beyond November, Adélie penguin colonies do not exist. This is so even if ice-free terrain, with a lot of pebbles for nests, is present. Fast ice or large ice floes (those wider than small towns) are also an impediment to Adélie penguins simply because the penguins cannot hold their breath long enough to exploit much of the water beneath. They might be able to reach 100 meters beyond the ice edge, but then they have to return to ice-free water to surface and breathe.

Finally, we must consider in more detail the annual cycle of sea ice in the Southern Ocean, which has important bearing on an Adélie penguin's life. The fact that there is an annual cycle at all is just as important to Adélie penguins as the actual existence of the sea ice itself. In the Arctic, even without the threat of polar bears, Adélie penguins probably could not exist, at least not in the numbers attained in the Antarctic. The Arctic Ocean is always covered by ice in a roughly continuous sheet, except for a fairly narrow zone of pack ice around its edges during winter in some very confined areas, such as the Bering, Labrador, and Greenland seas. A huge proportion of the Arctic shoreline is icebound for most of the year (although this is changing with global warming). Only seabirds that can fly in the air (the murres [*Uria* spp.], for example) are capable of moving between open-water feeding habitat and dry-land nesting terrain sometimes 100 or more kilometers away. Besides, so small are these pack-ice areas in the Arctic and so close are they to the peopled continents that the Adélie penguin populations would be small and isolated and, like those of the great auk, would have been exterminated by humans long ago.

When at its minimal extent, always in early February, pack ice in the Southern Ocean exists in just a few refugia, distributed from west to east: western Ross Sea, Amundsen Sea (major), southern Bellingshausen Sea, western Weddell Sea (major), and a series of much smaller areas along the coast from the Weddell eastward to the Ross Sea (fig. 2.2). This distribution of ice has important bearing on the location or migrations of the penguin's breeding populations because in February Adélie penguins must molt, which the majority do when positioned on a large ice floe for a few

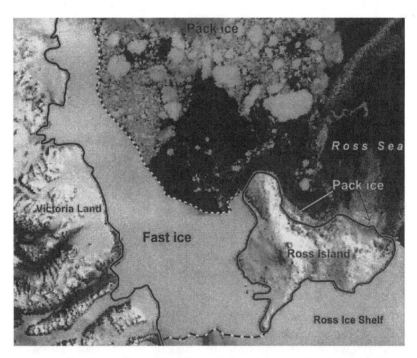

Figure 2.4 Fast ice covering much of McMurdo Sound, between Ross Island and Victoria Land.

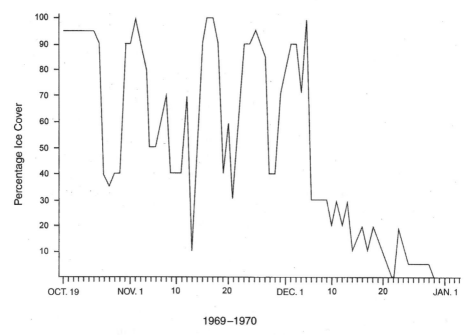

Figure 2.5 Seasonal changes in the proportion of the southern Ross Sea covered by sea ice in view from Cape Crozier, Ross Island (78°S). *Redrawn from Ainley and LeResche.*[4]

weeks. In late February the sea begins the freezing phase of its annual cycle, at least at highest latitudes,[118] at the northern edge of the sea-ice zone, such as the northern tip of the Antarctic Peninsula (South Orkney Islands, 61°S), freeze-up does not begin until June.[91] The growth of sea-ice extent is rapid from late March through June. Maximal extent is reached between July and September, usually in August, but throughout these three months the total ice extent usually is within 10 percent of the maximum. Therefore we can say that July through September is the period when the most sea ice exists in the Southern Ocean. Thereafter, it begins to recede, first at the northern periphery (it is gone around the South Orkney Islands by October[91]) and, finally, in the far south (disappears in southwestern Ross Sea by about January 1, although not always; fig. 2.5).

- **At-Sea Range and Habitat**

Adélie penguins live around the globe at its southern end, always associated with or in close proximity to pack ice. In other words, sea ice at its maximal extent defines the at-sea range of this species (fig. 2.1). The outer edge of the pack ice is the environmental boundary of the Adélie penguin's range. Certainly, during February, when the pack ice is minimal, Adélie penguins do live in what are then ice-free waters, particularly as they swim from nesting colonies to the pack ice refugia where they will molt. I have heard reports of large numbers of Adélie penguins seen in the vicinity of the Antarctic Polar Front, which is well north (hundreds of kilometers in some places) of the outer edge of the pack ice even in winter (figs. 2.1 and 2.2), but I find it hard to believe that they venture so far north. Although I have spent thousands of hours censusing seabirds in the Southern Ocean—in the South Pacific, South Atlantic, and the Ross, Amundsen, Bellingshausen, Weddell, and Scotia seas as well as the Drake Passage, mostly during the summer but also during fall and winter—rarely have I encountered an Adélie penguin in waters more than a day's sailing from sea ice and almost never beyond what is called winter water. The latter is the portion of the Southern Ocean surface layer with a salinity characteristic of an ocean where sea ice resided within the past year. This water is slightly fresher because of the melt of the ice (which is fresher than seawater because of the salt rejection in the ice-formation process).

The Antarctic Polar Front (formerly Antarctic Convergence) is the oceanographic boundary of the Antarctic or the agreed-upon northern boundary of the Southern Ocean. It lies generally between 55° and 60°S latitude, with some variation, principally in the Scotia Sea region where it juts northward (fig. 2.6). At the Polar Front, the colder, more dense surface water of the Southern Ocean, flowing north from the continent, meets the warmer, less dense surface water of the subantarctic (temperate) zone (fig. 2.7). Where the two waters converge, the colder Antarctic water sinks beneath the warmer. In some places, the meeting is so abrupt that flotsam, such as kelp pieces from the shallows around subantarctic islands, can be found floating, inhibited from further southward excursion by the north-flowing Antarctic waters. Human-made refuse, including plastics, accumulates there as well.

The waters south of the Polar Front flow east, propelled by incessant westerly winds. This flow has been called the Antarctic Circumpolar Current (ACC). The northern edge of the sea ice during winter coincides with the warmer waters at the southern boundary of the ACC.[70] South of the ACC, the general direction of flow is in the opposite direction, the East Wind Drift. The southern boundary of the ACC is important to

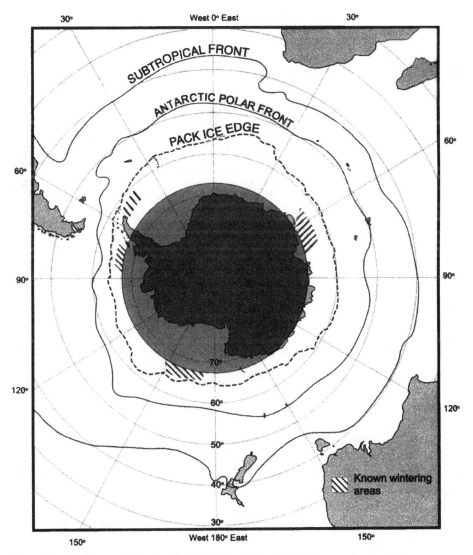

Figure 2.6 The large-scale factors that affect the choice of wintering areas by the Adélie penguin. Southern Ocean showing Antarctic Polar Front[64] (*solid line*), southern boundary of Antarctic Circumpolar Current[70] and pack-ice edge[39,118] (*short dashes*), and Antarctic Circle (*shading to the south*). Heavy cross-hatching shows where concentrations of Adélie penguins have been found during winter.

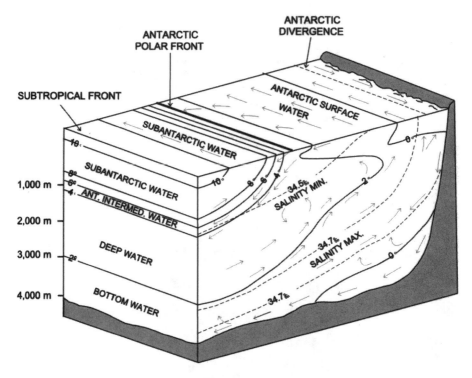

Figure 2.7 Schematic showing regions of surface water in the Southern Ocean from the Antarctic continent outward to the Antarctic Circumpolar Current (Antarctic Surface Water), Antarctic Polar Front, and beyond. *Redrawn from Knox.*[54]

Adélie penguins because waters to its south are especially rich in the organisms that the penguins eat. These waters are also important feeding grounds for baleen whales,[68,94] which follow the seasonal retreat of the pack ice. Farther south lies the continental shelf break, along which upwelling of Circumpolar Deep Water occurs (fig. 2.7). This process leads to even richer production of prey for the whales, penguins, and other seabirds.[2,3,49]

Another boundary important to Adélie penguins is the Antarctic Circle. The Antarctic Circle, well south of the Polar Front, is the celestial boundary of the Antarctic (fig. 2.6). It is at 66.5°S and marks the point where a line from the sun is perpendicular to the earth's surface at the austral summer solstice. South of that line, between the day of the spring solstice and the day of the autumn solstice, the sun does not set. The Adélie penguin ranges both north and south of the Antarctic Circle, at least during summer.

The ocean north of the pack ice and winter water is the habitat of the Adélie penguin's closest relative, the chinstrap penguin. The chinstrap looks just like a juvenile Adélie but with a very thin line around the throat. This "strap" is visible only at very close range, so it is very difficult to distinguish between the two species when tossing about on the sea. This uneasiness in distinguishing the two species has been expressed by biologists associated with early whaling expeditions.[18] Reports of Adélie penguins at Bouvet, an island well north of sea ice, proved to be chinstrap penguins upon subsequent inspection of photographs taken.[66] Any sightings of Adélies north of the pack ice and winter water probably are sightings of chinstraps.

Most penguin species are difficult to identify when seen porpoising at sea, especially from the large ships on which most of us travel in the Southern Ocean. This problem is exacerbated by the large and numerous waves, mist, and fog that are common in the Southern Ocean. The field guides show all kinds of seemingly gaudy head markings on many species that one would expect to see from a great distance. However, in my experience, except for the large species (king, emperor, yellow-eyed) and the small ones (little blue [*Eudyptula minor*]), which can be distinguished easily by their size, almost all other penguin species seen at sea look like an Adélie penguin: dark above, white below, with no face or head markings, and all about the same size. Even the crested penguins, which have the most gaudy markings of all, have their crests sleeked at sea and are difficult to identify. Even the big red bill of a crested male that is porpoising (i.e., surfacing to breathe) is difficult to see. The crested penguins (five species) and chinstrap penguins together outnumber the rest of the world's penguins combined, and these are the penguin species found in waters between the ice pack and the Antarctic Polar Front.

WINTER

A very detailed, careful study of the avifauna of the pack ice edge in the southern Scotia and northern Weddell seas in all seasons[10,11] showed that given a choice (i.e., not having to reach a coastal, land-based nest, as it has to during spring), the chinstrap penguin avoided opportunities to venture more than a few dozen kilometers into the pack ice. Even then, it did so only where ice floes were widely scattered (fig. 2.8). This avoidance was especially obvious during winter, when it was not necessary for penguins to visit nesting colonies and therefore cross habitat that otherwise they would not. In turn, Adélie penguins were not found north of the pack ice during winter. In fact, they occurred well south of the pack ice edge (and south of the southern boundary of the Antarctic Circumpolar Current; fig. 2.6) where the sea was not subject to the repeated freeze-thaw cycles that result from changes in wind direction. These observations supported the qualitative observations of whalers and early explorers of the Antarctic region (summarized elsewhere[35,66,86]). Less extensive observations by more contemporary researchers support these observations as well.[51,55,90] In the Scotia-Weddell study, as successive low-pressure systems approached from the west (every eight or nine days or so), the wind blew from the north and was warm enough to melt the thin, mostly new ice a few hundred kilometers south. Then, the front passed and the wind switched abruptly to blow from the south. With the wind blowing across the ice pack from the continent, the air temperature plummeted and the sea froze outward hundreds of kilometers in a few hours. This was the no-penguin zone: Despite an abundance of food, that zone was far too unstable for the Adélie or the chinstrap, which unlike aerial seabirds cannot respond quickly to changed conditions. During this study, large numbers of Adélie penguins were found in the outer 100 kilometers or so of the stable portion of the pack ice but few were found in its interior (fig. 2.8).

A study that used satellite transmitters to track Adélie penguin movements also provides some clues about the whereabouts of these penguins in winter.[53] Transmitters were glued to the feathers of the backs of four Adélie penguins after the birds had completed their molt in February. The study location was off MacRobertson Land, which is in the Indian Ocean sector of the Antarctic. There the pack-ice zone is narrow because the shore of the Antarctic continent in that sector is positioned farther north than elsewhere (65–67°S), except for the Antarctic Peninsula and therefore is near the southern boundary of the Antarctic Circumpolar Current (and warmer water). After their molt, which was undertaken at a colony (not where most Adélie penguins molt; see ch. 4), these four birds labored north in the heavy pack and fast ice to ice overlying the edge of the continental slope. There, judging from satellite imagery, they found looser, more open pack ice created by the shearing action of the Antarctic Circumpolar Current (fig. 2.7). The latter flows from west to east north of the continental shelf break.[70] The penguins thereby found an area where reliable access to the sea was possible. The pack ice ex-

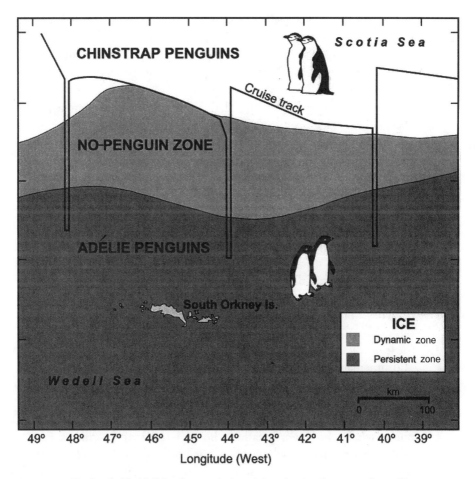

Figure 2.8 The Scotia-Weddell Confluence during winter, showing the areas where chinstrap and Adélie penguins were found and the no-penguin zone of unstable sea ice that lies between. *Redrawn from Ainley et al.*[11]

tended beyond that point, much farther north. However, the shelf break region has been found to be very rich in food.[2,3,49]

Complementing the winter study tracking the four Adélie penguins was a study that tracked a single chinstrap penguin.[110] A global location sensing instrument was attached to this penguin. Such an apparatus is smaller than a satellite tag, but researchers must recapture the penguin to retrieve the archived position data. This particular bird was recovered at its colony on King George Island six months after the instrument was attached. The record showed that during the winter, the chinstrap dwelt mostly in the open water north of the pack ice in the Scotia Sea or sometimes in the very loose ice at the pack-ice edge. That the chinstrap remained mostly in open water was not surprising, given the habitat preferences of Adélie and chinstrap penguins.

Additional observations shed light on the range and habitat preferences of the Adélie penguin. These occurred at about the same time of year as the satellite transmitter study but in the Amundsen Sea. I made a census of all seabirds, from the continent out beyond the ice pack edge, to determine where seabirds went with the approach of winter given the physical oceanographic and ice variables.[3] During this study, two important things took place. First, the Adélie penguins finished their annual molt. This is a process they must undergo every year to renew their feathers; while molting they usually remain on an ice floe for a few weeks. Second, soon after most Adélies had completed the molt, the sea surface began to freeze rapidly into a continuous sheet. This was the beginning of the freeze-up, indicating the start of winter. It happened in just a day or two, in one of the most ice-choked parts of the Southern Ocean. To this point, open water (leads) existed between ice floes, but with the freeze the leads disappeared.

Immediately after the freeze-up, I noted that minke whales, present in large numbers just the day before, had disappeared. As they swam north, they left holes in the new ice, and in these holes were crabeater seals (*Lobodon carcinophagus*) and emperor penguins. With this observation of minke whales and their breathing holes, I realized two things. First, I could see why the minke whale has such a long, hard "beak" (its Latin name is *Balaenoptera acutorostrata* [i.e., baleen whale with a very sharp beak]; the antarctic form recently was named *B. bonarensis*). Without the ability to punch holes in new ice, the whale would not be able to live as far south into the pack ice as it does at that time of year, nor would the seals and penguins that also use those holes. Second, I found that the pioneering experiments conducted on the diving physiology of emperor and Adélie penguins by Jerry Kooyman were not as cruel as they initially appeared.[56,58] Captured penguins were taken to an area of extensive fast ice. With no access to open water, they had to swim from one breathing hole, drilled artificially through the ice, to the next at increasing distances, enabling researchers to study their diving physiology. On the basis of my observations, it appears that pack-ice penguins are adept at this activity, and Kooyman did not really teach them a thing. Using holes made

Illustration 2.1 Minke whale, about 8 meters long, diving between icebergs. *Drawing by Lucia deLeiris © 1988.*

in the ice by the minke whale, the penguin can escape the interior of the pack ice during the initial freeze. The holes made by the whales were a few hundred meters apart, which is an easy breath-holding distance for the emperor but a stretch for the Adélie penguin.

Another thing that I noticed brought into focus some interesting observations on penguin navigation conducted in the 1960s by John Emlen and one of the pioneers of Adélie penguin behavioral research, Richard Penney.[34] Several Adélies were captured and taken to the polar plateau, an environment novel to the penguins, with no topographic features that penguins could use for navigation. Upon release, as long as the sun was visible, the penguins headed northeast in a straight line. Whenever clouds obscured the sun, the penguins meandered. Emlen and Penney concluded that the Adélie penguin used the sun for navigation. With the freeze-up I observed, the Adélie penguins disappeared, leaving only foot tracks or tobogganing tracks but *all* heading northeast across the new (and old) ice. By heading in that direction, the penguins could find open water or loose pack in the shortest time. By walking northeast, the penguins compensated for westward movement of the ice pack. Near the continent, the ice pack in general is moved west by the easterly winds in what is known as the East Wind Drift, south of the Antarctic Divergence (fig. 2.7).[54] The position fixes of the four Adélies to which the satellite transmitters had been attached, as described earlier,[53] indicated a westward movement in the pack ice. I wonder whether these penguins also were trying to compensate for the ice movement by walking northeast. It is also possible that the transmitters eventually detached from the penguins because the direction and rate of movement began to match that of the moving ice in that area after a time.[43]

Murray Levick also marveled at the navigation abilities of Adélie penguins and surmised that they found their way using means other than just sight:

> Duke of York Island, some twenty miles south of the Cape Adare rookery [where Levick was camped], another breeding-place has been made. This is a small colony only, as might be expected. Indeed, it is difficult to see why the penguins chose this place at all whilst room still exists at the bigger rookery, because Duke of York Island, until late in the season, is cut off from open water by many miles of sea-ice. . . . When the time arrived for the birds to feed, some open leads had formed about half way across the bay, and those of the Duke of York colony were to be seen streaming over the ice for many miles on their way between the water and their nests. They seem to think nothing of long journeys, however, as in the early season, when unbroken sea-ice intervened between the two rookeries, parties of penguins from Cape Adare actually used to march out and meet their Duke of York friends half way over. . . .
>
> To realize what this meant, we must remember that, an Adélie penguin's eyes being only about twelve inches above the ground when on the march, his horizon is only one mile distant. Thus from Cape Adare he could just see the top of the mountain on Duke of York Island peeping above the horizon on the clearest day. In anything like thick weather he could not see it at all, and probably he had never been there. So, in the first place, what was it that impelled him to go on this long journey to meet his friends, and when so impelled, what instinct pointed out the way? . . . In the case of the penguin, its horizon is so very short that it is quite evident he possesses a special sense of direction. (pp. 11–12)[60]

The general picture during winter, then, is that Adélie penguins live in the outer portion of the pack ice but not at its outermost edge. There they find loose but stable concentrations of ice floes, with open water in sufficient supply to allow easy access to the sea. In other words, the Adélie penguins seem to stay where the ice cover ranges between 15 and 80 percent. At the lower latitudes of the outer pack, Adélies also find another important resource: daylight. Besides needing the sun to navigate, Adélie penguins (and other penguin species) are visual predators, and it has been discovered that they usually feed only as deep as the penetration of light into the water allows (i.e., to about the 1 percent daylight level) (fig.

2.9).[109,113] Light penetration is impeded by water turbidity, which is extremely low during winter, but also by the steep angle of the sun's rays, especially in winter. Light is further attenuated in the shade of ice floes. During much of the austral winter, there is significant light only north of the Antarctic Circle. I hypothesize that most Adélie penguins are found north of the Antarctic Circle in winter[37] but south of the southern boundary of the Antarctic Circumpolar Current (fig. 2.2).

Few winter data are available to evaluate this idea. In the winter study at the pack ice margin in the Scotia-Weddell Confluence, there were plenty of Adélie penguins in the study area, which extended to about 62°S (well north of the Antarctic Circle). On the basis of the stage of digestion of prey taken by the Adélies, foraging occurred during the short period of daylight each day.[9] In a study using satellite transmitters in the southern Indian Ocean sector during winter, penguins also remained north of the Antarctic Circle.[53] Only at breeding sites of the Antarctic Peninsula, which are well north of the Antarctic Circle (e.g., Anvers Island) do Adélies visit land during the winter.[71]

With the coming of spring, Adélie penguins begin to migrate from the outer reaches of the ice pack toward land, where they will breed. This movement has rarely been observed directly because the pack ice is heaviest at that time of year, making travel by ships difficult. However, observers on the *Aurora*, which in 1915–16 was icebound in the outer pack ice north of the continent, confirm such a migration:

> During the first three months of the drift, which began from Cape Evans [78°S] in May, no Adélie penguins were seen or heard. The first five were seen on 5th August, the day on which the sun should first have been visible from the ship, had it not been for heavy drift. These birds, and fresh tracks of others, came from the northward across the pack, and more were seen daily making a gradual southward trek with the reappearance of the sun. During September the Adélies seen were passing southward at a greater speed, but not in great numbers. Their progress was at times stopped when they had to "lie-up" during blizzards. The ship's position was about 90 miles south of Sturge Island on 15th September [Balleny Islands, ca. 66°S]. Parties of adults continued to pass by throughout October. (p. 65)[35]

Much of this travel is undertaken by walking or tobogganing over the ice or by swimming in polynyas. The latter, a Russian-rooted word, refers to areas of open water or loosely spaced floes within the otherwise con-

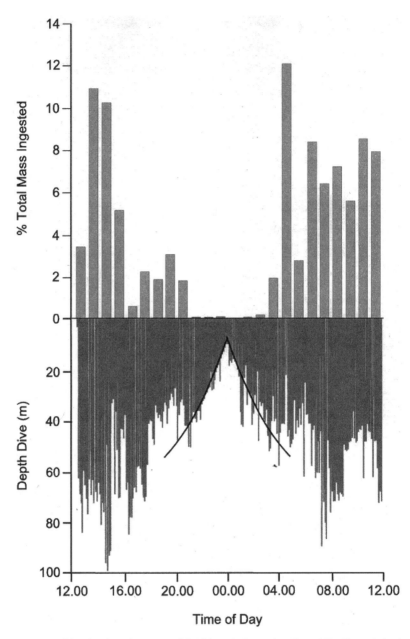

Figure 2.9 Dive depth and amount of food ingested as a function of the time of day in early January (62°S). Dive depth was determined using time depth recorders; food ingested was estimated using sensors that measured stomach temperatures (a drop in temperature occurred when food entered the stomach). *Redrawn from Wilson et al.*[113]

solidated ice pack. These areas usually are kept open by strong, persistent winds or upwelling of warm water. In fact, many Adélie penguin colonies exist where polynyas occur, but this is a subject for another chapter (see ch. 3). The over-ice migration is undertaken only after the penguins have accumulated a huge supply of body fat, which sustains them while the southern portion of the ice pack is too consolidated to allow reliable access to the sea and its food supply. Throughout the summer, as the ice pack dissipates, Adélies seek ever diminishing areas of pack ice in which to feed. In areas where the annual presence of pack ice has all but disappeared (western coast of the Antarctic Peninsula), so has the Adélie penguin, as discussed in chapter 9.

SUMMER

The summer distribution of Adélie penguins is much better known. The most intensive study spanned the entire Ross Sea during December and January.[7] As discussed in more detail in chapter 3, about 30 percent of the world's Adélies breed on islands and coasts bordering the Ross Sea. Therefore this particular study is important in understanding the at-sea distribution of this species in summer. Results showed some interesting patterns. First, the Adélie penguins were found where there was loose pack ice but not in extensive areas of open water that early in the study (December) were covered by ice (fig. 2.10). However, petrels and marine mammals also avoided this large open water area of the central Ross Sea, a curious phenomenon that to this day warrants further investigation.[2,7] Obviously, food availability had changed. Second, all Adélie penguins, even yearlings (which can be identified by their white chins and rarely visit breeding colonies), were concentrated in the pack ice within 150 kilometers of breeding colonies (fig. 2.11). The yearlings remained the farthest away. What led the nonbreeders to move toward but not all the way to breeding colonies is unknown. They may have followed the breeders from the wintering grounds, or their reproductive hormones were in sufficient quantity to instill an incipient migration toward breeding colonies. The fact that juveniles are found more distant from colonies than presumed adults has been commented on by many observers, especially from early expeditions adrift in the ice pack.[18,35,79,97] Finally, the southern portion of the Adélie penguin population in the Ross Sea (about one-third of the total population) was occupying waters overlying the continental shelf; the northern portion (two-thirds) was occupying waters over the continental slope. Continental slope waters around the Antarctic continent are particularly rich in food, and many marine birds and mammals are found there.[2,3,49]

Large concentrations of Adélie penguins in the pack ice offshore of breeding colonies have been observed in a number of other locations. This

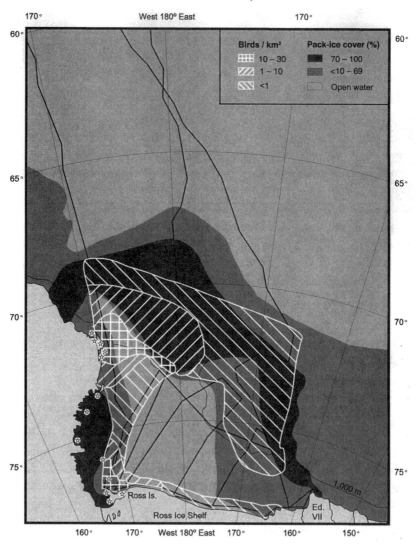

Figure 2.10 The distribution of Adélie penguins in the Ross Sea during summer in relation to the occurrence of sea ice. *Redrawn from Ainley et al.*[7]

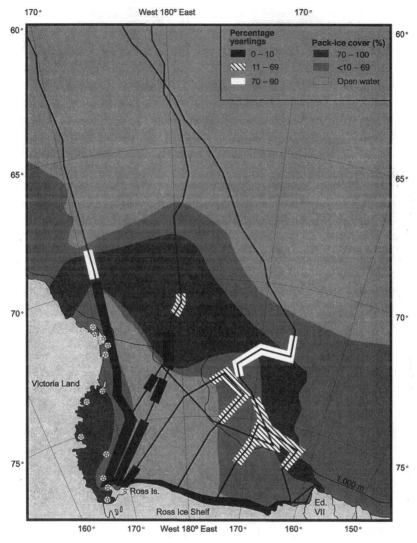

Figure 2.11 The geographic distribution of yearlings as a proportion of all Adélie penguins present in the Ross Sea during summer. *Redrawn from Ainley et al.*[7]

pattern was evident in the Scotia-Weddell Confluence region.[10] There, large colonies of this species occur in the South Orkney Islands and along the northeastern coast of the tip of the Antarctic Peninsula. With distance westward (toward the peninsula) from the northern end of a north-south center line of the Weddell Sea, Adélie penguin densities increased, especially in the pack ice between the South Orkneys and the Antarctic Peninsula. The latter area includes waters of the continental slope; immediately offshore of the colonies on the eastern shore of the peninsula would be the continental shelf, but the Adélie densities as a function of distance offshore of these colonies have not been investigated. High densities of Adélie penguins also have been observed in the area around the South Orkneys.[51] In waters off the western shore of the Antarctic Peninsula, in the Drake Passage and Bransfield Strait, Adélies were seen in "Weddell Sea water," which had flowed westward, but were absent in other water types.[48] Essentially, Weddell Sea water is winter water, as described earlier.

In the Indian Ocean sector of the Antarctic, in the pack ice offshore of MacRobertson Land, studies using telemetry and satellite imagery have shown that breeding Adélie penguins remain in the pack ice within 100 kilometers of the colonies.[53] Many Adélie penguins in that sector frequent the pack ice overlying the rich waters of the continental slope, as do the Adélies of the northwestern Ross Sea; others forage over the continental shelf, as do the Adélies of the southwestern Ross Sea. Offshore of MacRobertson Land, the proportion of the population foraging in depth-defined habitat is not known. Other findings indicate that most of the population in this region remains over the shelf during summer, in part because deeper waters are devoid of pack ice (fig. 2.12).[63,114]

WINTER AREAS IN RELATION TO SUMMER AREAS

In the central Ross Sea during late January, large numbers of Adélie penguins have been observed porpoising in northeastward transit from the southwest (where large colonies exist), across open waters, to the pack ice of the Amundsen Sea off Marie Byrd Land.[7] High densities of Adélies were then found in the pack ice that extended northwest from the Antarctic coast of Marie Byrd Land (fig. 2.13). Subsequently, observations in mid-February to March found high densities of Adélie penguins molting on the pack ice of the western Amundsen Sea.[3] Only very small Adélie penguin colonies exist in that region, so the breeding colonies of these birds probably include those of the southern Ross Sea. On the other hand, satellite transmitters placed on two birds that had just molted at

Figure 2.12 (opposite) Distribution of the Adélie penguin in the Indian Ocean during November and January-February. *Redrawn from Woehler et al.*[116] Ice position drawn from Gloersen et al.[39]

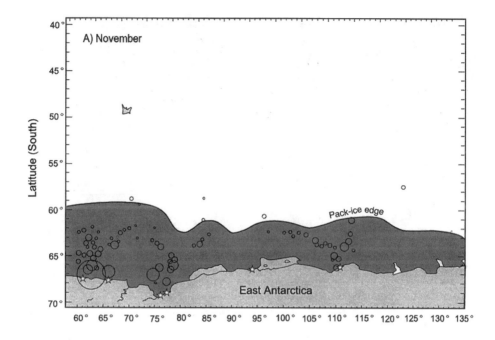
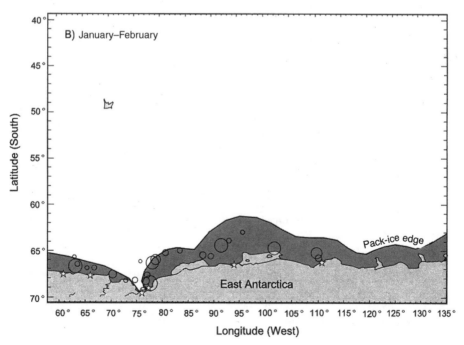

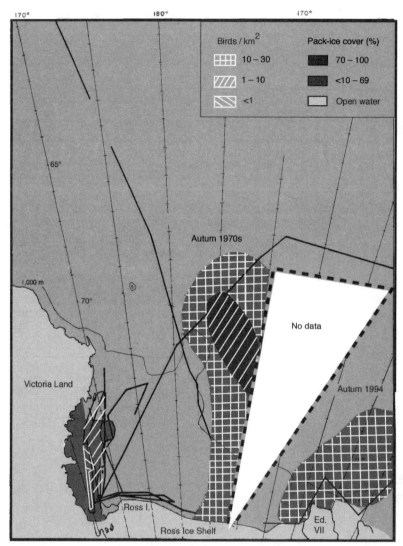

Figure 2.13 Distribution of Adélie penguins in the Ross Sea and Amundsen Sea sector during early autumn. *Data from Ainley et al.*[3,7]

Cape Bird in the southwestern Ross Sea moved northward along the Victoria Land coast to a small area of persistent pack ice near the Balleny Islands (fig. 2.13).[31] How representative these two birds were of Ross Island breeders—whether the huge satellite packs allowed them to behave normally or whether these birds were part of the southern Ross Sea breeding population at all (or nonbreeders visiting from the north coast of Victoria Land)—is unknown. Most Adélie penguins of high latitude do not molt at the colony where they breed.

As the Amundsen Sea census progressed eastward, passing through areas offshore of a coast too icebound to encourage large breeding colonies of Adélie penguins, densities of Adélie penguins declined sharply. The species did not appear in large numbers again until the census reached the southeastern Bellingshausen Sea (fully 100° farther east along the Antarctic coast, or a third of the way around Antarctica), where larger breeding populations of Adélie penguins exist. In fact, the Adélie penguins that nest along the western coast of the Antarctic Peninsula northward to about Anvers Island (64°S) all winter to the south in the pack ice of the southeastern Bellingshausen Sea.[37] Those on the northern and eastern coast of the Antarctic Peninsula apparently move east to the pack ice of the Scotia—Weddell Confluence.

The relationship between breeding and wintering areas is not well known for other coasts of Antarctica. The study using satellite imagery off the coast of MacRobertson Land indicated that Adélies breeding on that coast may winter directly offshore in the pack ice.[53]

■ Diet

Numerous authors describe the Adélie penguin as a krill predator, also inferring that the krill involved is the Antarctic krill (*Euphausia superba*). This classification is based in part on fact and in part on biopolitics. Beginning with the *Discovery* expeditions of the 1930s, which mark the beginning of the current age of marine research in the Southern Ocean (supplanting the heroic age of Cook, Weddell, and Bellingshausen),[40] no organism has received anywhere near the research effort that has been directed toward the Antarctic krill. First, this effort was justified by the fact that these crustaceans were the primary food of whales, with whaling an important commercial venture, and to understand the prey was to understand the predator. Then, with the overexploitation of whales, the krill research effort was justified by the belief that a krill surplus developed (because the cetacean predators had disappeared), which then could be exploited directly by trawl fishing. In the early 1980s, a fishery treaty was signed: the Convention for the Conservation of Antarctic Marine Living Resources (CCAMLR), put forth initially by nonfishing nations to en-

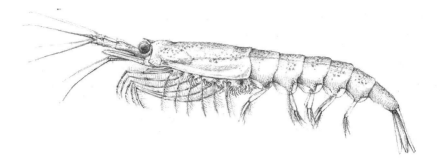

Illustration 2.2 As adults, Antarctic krill reach 5-6 centimeters in length. *Drawing by Lucia deLeiris © 1988.*

sure that if a krill fishery were developed, enough krill would remain for the whales to recover.[44] CCAMLR also covers finfish fisheries, especially those around subantarctic and Antarctic islands, but these stocks were depleted before CCAMLR took effect. CCAMLR includes a landmark clause that states that fish and krill stocks should not be exploited to the point that the ecosystem has no chance of recovery. With that clause came the CCAMLR Ecosystem Monitoring Program and a hope that Southern Ocean fisheries could be managed using an ecosystem perspective.

Throughout this period, if you were studying an organism that competed with whales for krill or might compete with humans for krill, the research was deemed relevant to society's problems. Therefore Adélie penguin research often was touted as krill predator research. This interpretation of research history contains a bit of cynicism, but given the facts of Adélie penguin diet, how else can we interpret the "krill predator" label? Labeling the Adélie as a krill predator has led some researchers to force square pegs into round holes in trying to make their data fit the accepted paradigm.

WINTER

Only one project to date has investigated the diet of Adélie penguins during winter.[9] This work was conducted in the pack ice of the northwestern Weddell Sea at about 61°S and to the east of the South Orkney Islands. Recall that this is the region where the breeding populations of the northern and eastern coasts of the Antarctic Peninsula and South Orkney Islands winter. Stomach loads were small (less than 200 grams) and were composed of squid, krill (*E. superba*), and fish, in order of importance by mass (table 2.1). The fish species eaten were those found only in waters overlying the deep ocean, that is, species of the family Myctophidae (lantern fish). Apparently the Adélies remained on ice floes, and when they were hungry and it was light enough to see, they dived off the floe to forage. In other words, that they did not seem to move around much, at least not laterally. On the basis of prey sampling done in conjunction with this study, penguins found all the prey they needed just underneath the ice floes.[9]

SUMMER

Almost all of our information on the diet of Adélie penguins has been gathered during a small portion of the summer: the chick-feeding period. William Emison[33] was the first researcher to use a pump to obtain the contents of a penguin's stomach, without having to kill the bird. Fifteen years later, Rory Wilson[104] perfected this technique, making an African penguin sick to its stomach by gently pumping a lot of extra water into it so that the penguin regurgitated its load. Today, this "water off-loading" technique is used by many researchers on most penguin species and is part of the CCAMLR Ecosystem Monitoring Program.

The first data on Adélie penguin diet were gathered by people needing to kill penguins to provide food for themselves or their sled dogs.[35,86,103] The reports through the 1950s were qualitative; for example, such descriptions as "almost all euphausiids" were evident. Most reports, with the majority coming from the accessible northern parts of Antarctica (i.e., the Antarctic Peninsula), indicated that euphausiids were the main or exclusive prey (table 2.1).

The first quantitative and one of the most well-rounded data sets on Adélie penguin diet (perhaps the most well-rounded description for any penguin except the African penguin), a combination of several efforts, comes from Ross Island in the southern Ross Sea. Recall that penguins in this area forage in ice-covered waters overlying the continental shelf. At Cape Crozier, Emison[33] collected fifteen complete samples in one year and sixteen the next between December 1 and January 31. Included were samples from breeders still incubating eggs (late stage of incubation), those with young chicks (still being brooded), and other parents with chicks old enough to be left unattended in the crèche. Full-stomach samples of incubating birds reached only about 200 grams (similar to winter loads), compared with 700–1,100 grams by the time the chicks entered

Table 2.1 Summary of information on the diet of Adélie penguins.

Location		Habitat	Diet	Reference
Antarctic Peninsula				
S Orkney Is (45°W)		Slope	Euphausiids	86
			E. superba, 100%	101
			E. superba, some fish	62
S Shetland Is (60°W)		Slope	*E. superba*, 99%	96
East coast (50°W)		Slope	Euphausiids	86
			Euphausiids	Bellingshausen in 77
			Euphausiids mostly	108
West coast, Anvers Is (64°W)		Slope	*E. superba* and nototheniid fish	Eklund in 77
			E. superba, *T. frigida*, and nototheniids	36
East Antarctica				
Hukuro Cove (39°E)		Slope	*E superba*, 58–64%	99
			E. crystallorophias, 2–4%	
			P. antarcticum and other fish, 26–34%	
Cape Hinode (42°E)		Slope	*E superba*, 79%	100
			E. crystallorophias, 10%	
			P. antarcticum and other fish, 4%	
MacRobertson Land	(63°E)	Shelf/slope	*E. superba*, 100%, slope	53
			P. antarcticum, 100%, shelf	
	(66°E)	Shelf/slope	*E. superba*, 39.7%, slope	52
			E. crystallorophias, 8.5%, shelf	
			P. antarcticum and other fish, 41.9%	
			Amphipods and squid, 3%	
Prydz Bay (77°E)		Shelf/slope	*E. superba*, 24%	77
			E. crystallorophias, 42%	
			P. antarcticum and other fish, 32%	
			E. superba, 85%	99
			E. crystallorophias, 7%	
			P. antarcticum and other fish, 8%	

Table 2.1 Continued

Location	Habitat	Diet	Reference
Wilkes Land (110°E)	Shelf/slope	Euphausiids	74
		E. superba, E. crystallorophias, and fish	25
		P. antarcticum and fish, 26–30%	102
		E. crystallorophias, 51–53%	
Adélie Land (140°E)	Shelf/slope	E. superba, 27–38%	102
		E. crystallorophias, 22–39%	
		P. antarcticum, 26–30%	
Victoria Land (170°E)	Cape Adare	Slope	Mostly E. superba	61, 103
	Cape Hallett	Shelf/slope	Euphausiids, 97%	Logan in 77
			Fish, 2%	
			E. superba and E. crystallorophias; fish late in season	Kinsky and Ensor in 77
	Franklin Is		E. crystallorophias, 99%	33
	Beaufort Is	Shelf	E. crystallorophias, 64%	33
			P. antarcticum, 32%	
	Ross Is	Shelf		
	Cape Royds		Fish and euphausiids	103
			P. antarcticum, 75%,	12
			E. crystallorophias, 25%	
	Cape Bird		E. crystallorophias, 99%	95
			P. antarcticum, 75%	12
			E. crystallorophias, 25%	
	Cape Crozier		E. crystallorophias, 60%	33
			P. antarcticum, 39%	
			E. crystallorophias, 25%	12
			P. antarcticum, 75%	

Open Ocean

Location	Habitat	Diet	Reference
Ross Sea, summer	Slope	E. superba, 98%	7
	Shelf	E. crystallorophias, 99%	
Weddell Sea, winter	Deep ocean	E. superba, 28%	9
		Myctophid fish, 14%	
		Squid, 54%	
Indian Ocean, summer	Deep ocean	Euphausiids, some fish	35

the crèche. Stomach loads reached the latter level shortly after the chicks hatched. At that time, perhaps in response to hormones, the Adélie penguin's stomach wall takes on an elastic character that allows it to expand dramatically. As it expands, the wall becomes paper-thin. Before then and during the winter, the stomach wall is tough and thick, and the stomach's capacity appears to be much lower.

The composition of the Adélie penguin diet, as observed by Emison, was composed of krill, but in this case crystal krill (*Euphausia crystallorophias*), 60 percent by mass, and Antarctic silverfish (*Pleuragramma antarcticum*), 39 percent. The fish became more important in the diet as the chick period progressed. On the basis of data gathered by the *Discovery* expeditions and later confirmed in more detail,[46] these two prey items are the principal middepth (i.e., not benthic, or bottom dwelling), micronektonic denizens in waters overlying the Antarctic continental shelves. The term *micronekton* includes organisms larger than zooplankton (organisms too small and weak to withstand the flow of ocean currents) but smaller than the penguins (nekton)—organisms of a size suitable for penguin food. Micronekton can swim against currents; in more familiar parts of the world they include small fish such as anchovies and herring.

Illustration 2.3 Antarctic silverfish adult, about 20 centimeters long—a bit too large for an Adélie penguin to eat. *Drawing by Lucia deLeiris © 1988.*

Illustration 2.4 Crystal krill and some copepods at the undersurface of an ice floe. *Drawing by Lucia deLeiris © 1988.*

Twenty years after Emison's work, at Cape Bird (about 70 kilometers from Cape Crozier) Yolanda van Heezik[95] pumped the stomachs of Adélies that were incubating eggs but much earlier in the cycle than studied by Emison. At this time of year, much pack ice still exists near the Ross Island breeding colonies. Stomach loads were small and were dominated by *E. crystallorophias,* with a few amphipods and few fish.

Most recently, samples collected throughout the chick-provisioning period simultaneously at three colonies on Ross Island indicate that the Adélie diet varies spatially and temporally in its mix of *E. crystallorophias* and *P. antarcticum.*[12] Where there was extensive pack ice, over all three study years, krill made up 50–90 percent of the diet. In contrast, where the habitat consisted of open water adjacent to fast ice, fish contributed 50–90 percent to the diet. The lowest percentage of fish was found in the year of extensive ice, and the prevalence of fish increased with time each year as the pack ice dissipated in the study area.

The most detailed, single-season record of diet was obtained at Magnetic Island, Prydz Bay, in the Indian Ocean sector.[77] Weekly sampling was conducted during the latter half of breeding in one season (1982–83) and the earlier part in the next (1983–84), for a total of 4.5 months of sampling. During incubation, *E. crystallorophias* and amphipods of several species predominated. The euphausiid remained important through hatching. As the chick period progressed, fish (especially *P. antarcticum*) increased in prevalence, becoming almost the exclusive prey by the time the chicks fledged. *E. superba* appeared during the crèche period but never contributed more than about 20 percent by mass. This contrasts with another study conducted at Magnetic Island in the early 1990s, when during the chick provisioning period *E. superba* contributed 85 percent by mass.[99]

Contrast these findings with stomach pump samples obtained at the South Orkney Islands, off the northeast tip of the Antarctic Peninsula, and at the South Shetland Islands, west of the tip of the Antarctic Peninsula.[62,96] The Adélie penguins were feeding in deep waters overlying the continental slope. During the chick-provisioning phase (the only portion of the penguin cycle sampled here) the diet was more than 95 percent *E. superba,* with a smattering of fish of unknown identity. Several less intensive studies, in which sampling occurred over the deep ocean or continental slope revealed that *E. superba* was the predominant prey (table 2.1). Accordingly, several diet samples taken from Adélies at sea during late December in pack ice-covered waters overlying the continental slope of the Ross Sea also were composed entirely of *E. superba;* the samples collected in the pack ice overlying the shelf were composed of *E. crystallorophias.*[7]

The remaining studies of the summertime Adélie penguin diet yielded results that are a combination of the continental slope and continental shelf results reviewed earlier. These studies were conducted at sites along the coast of the continent in the Indian Ocean sector of Antarctica. In these cases, penguin foraging over the continental shelf was limited by the presence of fast ice. In some cases, the penguins dove through tide cracks near the shore (as noted earlier in Levick's observations); in others, they trekked across the ice to looser pack overlying the continental slope. This was possible because the continental shelf in this sector is less than 100 kilometers wide. First, at Hukuro Cove, the diet of Adélies that dove through tide cracks during three different years was 58–72 percent *E. superba.*[99] Throughout the sampling period, much sea ice was present. The remainder of the diet was composed of fish, either *P. antarcticum* or some ice-associated species also found over the continental shelf (*Trematomus* spp.). Farther to the east along this coast, a limited amount of sampling was conducted of incubating and chick-rearing Adélie penguins nesting at Béchervaise Island, MacRobertson Land.[53] In this case, however, birds whose stomachs were pumped had previously been fitted with satellite transmitters, so the actual foraging area was known. Adélie penguins making long trips to forage at the shelf break ate a lot of *E. superba,* but those feeding over the continental shelf ate *P. antarcticum* and *E. crystallorophias.*

As indicated in the studies at several colonies on Ross Island, the temporal and spatial variation in diet is also evident in studies taking place in East Antarctica. This is especially true in diet studies conducted at sites in Wilkes Land and nearby Adélie Land.[52,102] There, as at Béchervaise Island farther west, the diet varied in composition depending on whether the penguins foraged over the shelf or at the shelf break.

To summarize, then, the Adélie penguin seems to eat a lot of fish, particularly during the late chick-provisioning period. Usually the fish species consumed is *P. antarcticum.* During winter the Adélie eats much squid and fish. Whether it can be called a krill predator depends on whether the birds are feeding over the continental slope or deep ocean. When feeding over the shelf, the penguin mixes fish with *E. crystallorophias,* more so when there is little pack ice, as occurs late in the breeding season. When the penguin feeds over deep water, the krill eaten is *E. superba,* and the fish is likely to be a myctophid rather than a nototheniid (*P. antarcticum*). On the basis of this record, I would not classify the Adélie penguin as a krill predator. Like almost all other seabirds, the Adélie is a dietary generalist.

■ Prey Capture

Penguins' prey species have devised various ways to avoid capture. The remarks of William and Peggy Hamner[41,42] are particularly telling in this regard, especially in light of estimates (based on energetic concerns) of how much krill an Adélie penguin must capture to maintain its energy and to feed its chicks. These authors warned against the assumption that a predator feeds at will and that all that remains is to measure feeding rate. Furthermore, they noted that predators do not kill their prey all that easily; if they did, the predators would soon go hungry. Selection pressure for predatory avoidance is always stronger than for successful capture of prey because the predator can afford to make mistakes whereas the prey cannot:

> Each species of krill predator finds and feeds on krill in unique species-specific was, whereas krill always live within schools and they must rely

on schooling as a generalized antipredator behavior that deters, on evolutionary average, all of the predators that eat krill. Krill do exhibit some unique and complex antipredatory escape behaviors in response to tactile predators [ctenophores]. . . . Krill predators can be broadly categorized into those that capture krill individually and those that are batch feeders. Penguins feed on isolated krill one at a time, whereas baleen whales are of necessity batch feeders that cannot be sustained by thinly distributed resources. (p. 199)[42]

The schools of krill and fish encountered by Adélie penguins also contribute to the penguins' success. Only when a school is located can the search for prey cease, at least temporarily. This is important for penguins, which, unlike aerial birds or whales, cannot search widely for prey. In shifting from prey search to prey capture, however, the penguin must adapt to the predator avoidance behavior of the prey.

The primary reason for schooling behavior appears to be predator avoidance[72] through the principle of safety in numbers ("the other guy will get eaten, not me") and the ability to startle the predator with an instantaneous, "flash" escape by all members of the school simultaneously. In chapter 7, we discuss how Adélie penguins use this behavior (flocking, scattering when chased) to avoid being eaten by seals.

Illustration 2.5 A school of Antarctic krill, all of similar size and all pointing in the same direction. *Drawing by Lucia deLeiris © 1988.*

A school (or flock) is an aggregation of individuals with polarized orientation, all moving in the same way at the same time.[73] Schools respond fluidly as a cohesive whole, so when a school moves to avoid a predator, individuals also have to avoid one another by obeying a set of rules that are beyond the scope of this discussion. This behavior protects individuals because stragglers or aberrantly behaving individuals are easily sighted and become the focus of the predator.[42]

Schools of Antarctic silverfish have not been observed directly, but those of krill have.[42,82] The fish probably behave like anchovies (*Engraulis* spp.) and other schooling fish, which form even tighter balls when attacked by a predator. The response of krill and other schooling crustaceans to predators has been observed directly. The prey's response is graded. First, they condense, polarize, and swim away. Second, if the pursuit continues, they confuse the predator by splitting and then reforming (coordinated tail flipping by euphausiids). This is the flash expansion of the school as it accelerates out of the path of attack. Finally, the crustaceans exhibit uncoordinated, vigorous tail-flipping, which breaks down group cohesion. Then the predator can take only one at a time. Many will escape.

How Adélie penguins adapt to the behavior of prey in schools is not known. What is known is that Adélie penguins forage in small flocks of up to ten individuals. Such flocks are typical during winter for this species and are much smaller than those that leave the breeding colonies on foraging trips. The latter flocks are composed of hundreds of penguins at times. These flocks are large on departure, probably to foil predatory leopard seals lurking along the beach. The flocks then divide into smaller groups, seemingly for more efficient foraging (and perhaps for continued predator vigilance).

During observations of Adélie penguins foraging from a tide crack in McMurdo Sound, Ross Sea, made by members of my research group in December 1999, it became apparent that they were foraging in concert. All individuals in the flocks of penguins dove simultaneously, in tight groups, then reappeared all at the same time one to two minutes later. On the other hand, observations made in Prydz Bay of penguins foraging near colonies in sparse pack ice indicated that the penguins foraged individually or in flocks composed of fewer than three individuals on average.[69] It seems that there is no rule governing whether penguins forage individually or in groups, or at least we don't know enough yet to perceive any rules. Actual prey capture by Adélie penguins has been observed, or at least reported, only once, by Douglas Mawson:

I once watched for a quarter of an hour with very great interest several Adélie penguins feeding in the shallow waters of Cape Denison. It was an occasion of perfectly tranquil sea, and standing upon a high knob of ice overlooking the water I could see quite clearly the movements of the penguins under water as if looking through clear glass. The penguins swam within a foot of the bottom (it was only 4 or 5 feet deep in that little embayment), traveling at a high rate of speed and resembling torpedoes going through the water, their stream-line shape offering evidently very little resistance. The speed, however, was not too great to observe just what they were doing. As they traveled in a zig-zag course backwards and forwards and round the little bay without slackening pace, one could observe their heads darting from time to time to the left and to the right, evidently picking up euphausians as they moved. In fact, their necks were jerking out and their beaks going just about as fast as a barnyard fowl feeds on grain thrown on the floor. (cited in Falla, p. 74)[35]

It seems possible, and perhaps more likely, that the penguins were feeding on small (juvenile) fish. The habitat Mawson describes is not the usual habitat of krill. By the time Mawson observed the scene, apparently they were in the final, scatter mode of predator avoidance. A bay less than two meters deep is very atypical of the Adélie penguins' foraging habitat, which normally is the ocean hundreds of meters deep. In such conditions, it is not possible to trap the prey against the bottom, as was happening in this example. Perhaps group foraging aids in the trapping of the fish and krill captured.

Mawson's observations are consistent with some anatomical characteristics related to food capture by Adélie penguins.[117] Observing the structure of the bones and muscles of the head and neck, Richard Zusi concluded that they caught and swallowed prey individually. This is consistent with Rory Wilson's[105] observation of other penguin species, specifically African penguins capturing anchovies. The penguins swam rapidly around the fish school, then dashed in to make a capture. Because they were capturing fish (larger than a euphausiid), the birds paused briefly after each capture to swallow before proceeding to the next fish. Large, spiny prey such as a fish must be swallowed head first. Otherwise, the spines can become stuck in the penguins' throat. The mouths of penguins, including Adélies, are densely lined with backward-projecting, hard papillae, which help them hold squiggling prey as the penguin works it into a headfirst position before swallowing. Presumably, a euphausiid or tiny, juvenile fish can be swallowed with much greater ease than a larger fish (as seems the case in Mawson's example).

■ Foraging Behavior

The subject of the foraging behavior of the Adélie penguin includes information on its travel between the colony and feeding area, prey searching, and prey capture. Thanks to the cooperation of penguins and the development of microelectronic instrumentation, much is known about all three of these subjects. In fact, more is known about the diving of penguins than of any other diving bird.

Penguins are highly capable divers. Studies using models, wind tunnels, and water courses have shown that penguins have such low drag that even marine engineers are incredulous.[17] The measured drag of a live penguin is significantly lower than what is called the ideal spindle, the most efficient shape for an object that has to move through the water.[29] The low drag of penguins results from their body shape, which the penguin can alter to suit its purpose. The regular roughness of their scalelike feathers disrupts adhesion of the water to their body surface. Although penguins beat their wings (flippers) just like flying birds, unlike the latter, penguins can generate propulsion on both the downstroke and the upstroke because of the shape of their flippers. In the process, penguins apparently generate almost no turbulence, except in the immediate vicinity of their tail. With little turbulence, drag is further reduced. Adélie penguins have the longest tail of any penguin; in fact, at one time they were known as the long-tailed penguin. How or whether their tail contributes to their hydrodynamic capabilities is unknown. Adélie penguins seem to use the tail as an aid in steering. It might be that they must be able to steer more abruptly than other species to negotiate brash and blocks of ice tossed about by the sea, especially along landing beaches.

TRAVELING

When on ice, Adélie penguins travel by walking or, if the surface is soft, by tobogganing.[86,107] Their very long toenails aid them well in this activity. As tracked by satellite telemetry, Adélie penguins walk over ice at 1–2.5 kilometers per hour.[53] This estimate includes pauses for rest and other reasons. Timed over known distances directly, without pauses, Adélies walk 2.6–3.9 kilometers per hour.[89]

When traveling in the water, Adélie penguins move by porpoising. Many other penguin species do this. They swim underwater, a few meters deep, and periodically leap clear of the surface without slowing their pace. In fact, with the much lower resistance offered by air, their speed probably increases. At each leap, they take a breath. The overall speed of penguins in traveling mode (swimming with periodic leaps for breath) is slowest in the smallest species and fastest in the largest.[105] At their most efficient speed (that is, when the cost of transportation, as physiologists call it, is minimized), Adélie penguins swim 2.2 meters per second, or 7.9 kilometers per hour. This rate, measured in an enclosed, elongated chamber in which oxygen consumption was measured, is much faster than walking and much more energy efficient.[28,111] Such a speed is equivalent to the speed of sailing ships, but not the very fast clipper ships; as noted by Ross,

> We saw many seals, as we sailed along, basking on the ice, and several penguins; these curious birds actually followed our ships, answering the call of the sailors, who imitated their cry; and although they could not scramble over the ice so fast as our ships sailed past it, they made up for it when they got into the water, and we soon had quite a flock of them in our wake, playing about our vessel like so many porpoises. (1:178)[83]

When in danger from leopard seals, Adélies are even quicker but cannot sustain the pace indefinitely. They do seem to be able to sustain

their pace when porpoising at their most efficient rate. In a swim tank, their maximum speed measured was 4.4 meters per second (or 15.8 kilometers per hour).[28] If pressed, Adélie penguins probably can swim much faster.

When in the long-haul mode of porpoising, they may leap to breathe only every minute or so. When porpoising near a colony or near ice floes onto which they intend to jump, they come up for a breath every few seconds, perhaps not so much to breathe as to judge where they will jump. In other words, they porpoise along an ice foot, then turn underwater and leap to the place that appears to have easiest access. Usually, only when they are sure that no leopard seals are about do Adélie penguins pause at the surface to look around. Adélies can leap to about 3 meters, which they seem to be able to do with minimal approach. Apparently, they have incredible powers of acceleration when swimming.

On the basis of hypothetical considerations (e.g., metabolic rate), an Adélie penguin at its most efficient speed can swim about 175 meters before it uses up the oxygen stored in its lungs, blood, and muscles.[29] At that point it takes a breath if near the surface. In other words, it is still swimming aerobically. However, it can continue to swim anaerobically if need be.[76] In the swimming tank, Adélies exceeded their aerobic limits on 54 percent of their dives (limits calculated to be seventy-six seconds). What their limits are in nature—that is, to what degree they will test their lim-

Illustration 2.6   A flock of porpoising Adélie penguins.
Drawing by Lucia deLeiris © 1988.

its—is unknown. Adélie penguin foraging dives average 115 to 230 seconds,[32,69,84] which is longer than their aerobic dive limit.

FORAGING RANGE

Much interest and discussion center around the distance from colonies that nesting Adélie penguins travel when obtaining food for themselves and their chicks. Information on foraging range has been used in several ways. First, when seeking food, central place foragers (such as Adélie penguins) must come and go from a fixed location (that is, their breeding colony) for a period of their life cycle. Fish and certain cetaceans can hang out wherever their food happens to be or feed in one location and then tend to their young in a more benign location using stored resources. Penguins can't do this during breeding. If there are enough central place foragers within the foraging range of a colony, then they can deplete the supply of prey.[13,75,111] Some questions arise. How much food must be available within foraging range of a penguin colony to support sufficient chick production to maintain population stability?[6,38] And when two or more similar species nest at a given site—for instance, all three pygoscelid penguins (Adélie, chinstrap, and gentoo) at various locations on the west coast of the Antarctic Peninsula—how might each affect the other's foraging behavior or prey consumption?[93,105] The answers might reveal how various penguin species differentiated and evolved morphologically and behaviorally. First, however, it must be established that the species in question actually compete for food, which is possible only if food is limited and they forage in the same place, at the same time, and on the same prey stocks. Finally there is a more practical question: Is a given Adélie penguin colony or population in competition for food with fishing boats?[53] Again, it is necessary to know whether the penguins and the boats will be fishing for the same prey, in the same place, and at the same time.[45,78] The whole CCAMLR Ecosystem Monitoring Program is based on this question.

The foraging range of Adélie penguins has been determined directly in only a few studies. To do this, triangulation of locations (using radio telemetry) or acquisition of fixes (using satellite transmitters) is needed. A radiotelemetry study of Adélie penguins nesting at Point Thomas, King George Island (to the west of the northern Antarctic Peninsula), found a maximum foraging radius of 48 kilometers during the chick-feeding period.[92] Another study,[53] this one using satellite transmitters placed on a small number of Adélie penguins nesting at Béchervaise Island, MacRobertson Land, found that during the incubation and chick-provisioning period two foraging patterns were evident, with some birds feeding within 20 kilometers (over the continental shelf) and others feeding at about 100 kilometers (continental slope). The majority of the individuals feeding close to the colony were males, and most of those feeding far away were females.[23,24] In a third radiotelemetry study, Adélie penguins fed by diving through tide cracks in fast ice, but because they had to walk they limited their range to less than 5 kilometers from their colony.[98,99]

The most complete picture of foraging range can be developed by combining the results of three radiotelemetry studies conducted at three colonies on Ross Island and a fourth at nearby Beaufort Island in the southwestern Ross Sea. During the incubation period, when a member of each pair remained away alternately for several days or longer, most Adélie penguins from Cape Bird moved 23–63 kilometers away. During chick provisioning, when nest attendance periods normally are short, foraging ranges were shorter.[32,84] When radios were deployed on penguins in this area, this time at all four colonies simultaneously during the chick-provisioning period, it was found that parents from the three smaller

Figure 2.14 Overlap in foraging areas among Adélie penguins breeding at four adjacent colonies in the southern Ross Sea. The numbers within the symbols denote number of observations at that location. *From Ainley et al.*[8]

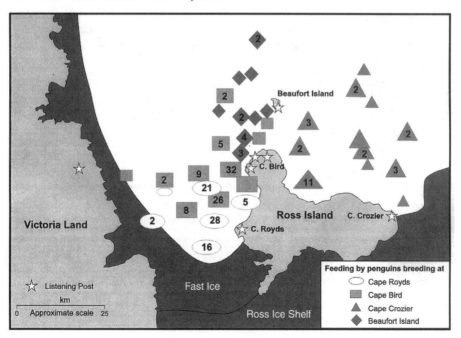

colonies fed within a radius of 30–40 kilometers and broadly overlapped their feeding areas. In contrast, parents from the largest colony fed out to 70 kilometers but without overlapping the other three (fig. 2.14).[81] When individuals repeatedly fed 70 kilometers away, their body mass decreased, the food loads delivered to chicks declined, and chick growth declined.[12] Therefore 70 kilometers may be the outer limit of the foraging range in this species when they are feeding large chicks. In any case, competition for food between the three smaller colonies was possible because foraging areas overlapped. Competition probably did not come about because the overall density of foraging birds was low.[8]

The remaining information on foraging range, all from sites along the coasts of the Antarctic Peninsula and outlying islands, was derived by extrapolating distance from duration of trips, the amount of diving that took place, and the swimming speed in transit. It was estimated that during the chick-provisioning period at Hope Bay, Adélies were feeding out to 14 kilometers;[111] at Arthur Harbor, Anvers Island, out to about 16 kilometers;[22] and at Ardley Island, within 20 kilometers.[105]

The foraging range of Adélie penguins differs from region to region as well as within and between years. On the basis of current information, there can be no clear-cut rule. Presumably, in all the studies just mentioned, parents were provisioning chicks adequately to produce viable fledglings. In other words, even with instruments attached and scientists fussing over them, the penguins were able to supply their chicks with food. In that respect, we know that some parents can sustain a foraging radius of 70 kilometers during chick provisioning,[81] but their chicks may not receive enough food, and we know that small transmitters (less than 1 percent the frontal cross section of the penguin) have almost no effect on the penguins.[16,99] Larger transmitters, such as satellite transmitters, appear to slow the penguins.[53]

At Béchervaise Island, East Antarctica, parents remain away longer to forage if their own body mass, upon departure, is lower than that of parents who stay away for a short period.[23] This is not to say necessarily that the foraging range was extended. Rather, the parents rebuilt their own energetic reserves before capturing prey for return to their chicks. Rebuilding took more time if they started out in deficit. This finding was consistent with the finding of the study at Ross Island in which parents provided for themselves before bringing food to their chicks.[12] Both studies indicate the dangers of extrapolating foraging range based on duration of foraging trips without having other information at hand.

The longer distance traveled by females on foraging trips, as detected in the study at Béchervaise Island,[23,24] was an interesting finding. Supposedly, the females were sacrificing frequent feeding of their chicks to travel farther (and longer) to exploit more reliable sources of prey. To do this they bypassed the neritic waters exploited by males to reach more productive waters of the shelf break front.[2,3,47] Apparently, they had sacrificed more of their body reserves earlier in nesting than did males, so their first priority was to replenish themselves. On the other hand, more distant foraging by females has also been observed in the southern Ross Sea, where the only available habitat for foraging is neritic waters.[12] Therefore much more information is needed on sexual differences in foraging behavior and capabilities.

FORAGING DIVES

Adélie penguins capture live prey by pursuing them underwater. The depth to which penguins can forage is a function of body size, with larger penguins able to dive deeper;[105] greater body size allows an animal ability to withstand the compression that comes with the pressures of greater

Figure 2.15 Data of the type obtained by time-depth recorders to show types of dives: A) porpoising, B) V-shaped, C–D) U-shaped or flat-bottomed dives. *Data from Wilson.*[105]

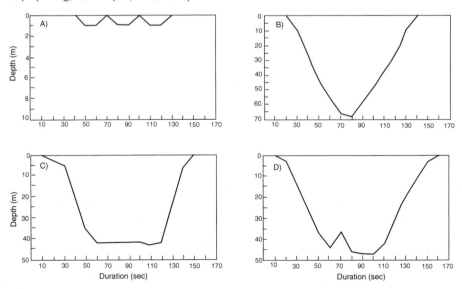

depth. Depth of foraging (not necessarily the same as depth of diving) is also limited by ambient light and its penetration into the water.[113]

Adélie penguins perform two dive types while foraging.[105] Once they arrive at a location, perhaps where they captured prey on a previous trip, they make a series of V-shaped or bounce dives, named for the shape of the time-depth recorder trace (fig. 2.15). In these dives, the penguins travel downward at a constant angle and speed, reach a depth, and then immediately return to the surface at a constant angle and speed. The angle becomes steeper for deeper dives. It is thought that such dives are made in search of a prey school. Because the angle and speed of descent are constant and are related to the final dive depth, it is thought that the penguin has some idea about how deep it is going to go before it starts. Once prey are located, the penguin's succeeding dives include extended periods when it remains at a certain depth. These are called U-shaped or flat-bottomed dives. Presumably, the flat portion of the trace indicates the depth where prey occur at a sufficient density to interest the penguin. Instruments attached to Adélies that record speed and direction of swimming indicate that when the penguin is at the foraging depth, its speed and horizontal direction change radically, presumably as it chases and captures its prey. The amount of time that a penguin can stay at depth is related to body size, with larger penguins remaining submerged longer.[105] As noted earlier, feeding dives average 115–230 seconds; the longest Adélie penguin dive recorded is 350 seconds.[69] Adélie penguins have been recorded to about 170 meters deep.

In discussions of the diving capabilities of penguins, much attention has been given to how deep they dive. Dive depth indicates certain things about a penguin's physiology, so depth is a critical parameter in describing diving abilities. On the other hand, it has been found that most of a penguin's feeding dives are much shallower than the maximum. For instance, most Adélie penguin dives are shallower than 50 meters.[105] In another respect, the great attention researchers give to depth of dive may be a holdover from the way in which instrumentation was developed. Initially, instruments sensitive to maximum pressure could record only maximum depth on any one (or succeeding) foraging trips. Then instruments were developed that could log the number of dives to prespecified depths. Finally, instruments (in use today) were developed to record the durations at which a penguin remained at any depth on every dive throughout a foraging trip. The entire dive chronology could be recorded on the computer chip contained in a time-depth recorder. This history of research is important to keep in mind because I suspect that it is more important to know how long they can remain submerged.

An important series of studies compared the diving of birds that were feeding in fast ice-covered waters with those feeding where there was pack ice.[98,99] Results were revealing. In waters loosely covered by pack ice, the penguins behaved as expected: maximum dive depth 119 meters, average dive depth 23 meters, and mean duration of dives 62 seconds. In marked contrast, however, was the penguins' performance when they were feeding under fast ice: maximum dive depth just 48 meters, mean depth 6–11 meters, and mean duration of dives 84–114 seconds.

Why are these results so revealing? They help us to understand why Adélie penguins do so well in ice-covered waters, which their close relatives (chinstraps, gentoos) try to avoid. The Adélie penguin can hold its breath longer than chinstraps.[105] Breath holding is more important than depth when exploiting prey hiding on the undersides of ice floes. Unlike its congeners, the Adélie penguin is a creature of the pack ice. Some biochemical studies of the oxygen-holding capabilities of the muscles of pygoscelid penguins indicate that the Adélie penguin's are superior to those of the chinstrap and gentoo.[14,15] In other words, the biochemical and physiological makeup of an Adélie predicts that it can swim aerobically much more readily than a gentoo or chinstrap penguin. After all, most of the oxygen penguins use in diving is stored in the blood and muscle tissue.[76] We now have a better idea of why this attribute developed. Recall that a portion of Adélie penguin prey hides in the channels and slush ice on the underside of sea ice. This is especially important during winter, when the only prey available are found either just under the ice, thus obtainable by Adélies, or very deep, thus obtainable by emperor penguins,[9] which can dive the deepest, to 535 meters at least.[105] The reason prey concentrate on the underside of the ice, especially during winter or in fast ice areas at any time of year, is that there is minimal penetration of daylight into the water. Therefore the Adélie penguin prey in this circumstance are captured as they graze on the algae that grow as close as possible to the strongest light, just under the ice or in the spaces between ice crystals. The 6- to 11-meter diving depths measured in Adélies feeding through tide cracks in multiyear fast ice probably are little deeper than the underside of the ice. Prey could be depleted quickly near a tide crack, because the crack is a very limited access point, so only by holding its breath longer to swim farther away from the crack would an Adélie penguin find additional prey.

The propensity (or need, especially during winter) to capture prey on the underside of ice floes perhaps explains another difference between Adélie penguins and their close relative, the chinstrap penguin. As mentioned earlier, the Adélie associates with ice and the chinstrap with open water. Accordingly, the chinstrap penguin, on average, can swim more rapidly than the Adélie.[112] Such an ability is necessary for a species that must search huge volumes of open water to find prey schools. The Adélie penguin can resort to hunting and pecking at prey lodged on the underside of ice floes, which entails slower swimming.

This discussion of capturing prey under fast ice leads us to the diel aspect of Adélie penguin diving and foraging. It has been observed repeatedly, but in studies conducted at low latitude (less than 70°S), that Adélies and other penguins dive to shallower depths during the darker part of the day than at midday. This pattern is related to the penguins' visual acuity at low light levels.[109,113] In the fast ice habitat described earlier (69°S), the penguins foraged only during midday.[98,99] Presumably this was because light penetration of the water was dramatically impeded by the ice, which besides a snow cover probably also contained a healthy growth of within-ice algae, which would block sunlight.[88] In the very high latitudes of the Ross Sea, there is no evidence that Adélie penguins are discouraged from foraging at the time of day during midsummer when the angle of the sun is lowest (midnight). In any case, at colonies there is no circadian signal in the numbers of penguins departing for or arriving from foraging trips.[30,65] Even at midnight there is more than enough light by which to forage. With decreasing latitude, however, the pattern is different. In areas of low latitude (i.e., less than 69°S), such as northern Antarctic Peninsula and Prydz Bay during the darkest hours even in midsummer, few Adélie penguins forage, and those that do dive only to shallow depths.[21,69,109] One study showed that Adélies captured 300 times more food per kilometer swum during the hours of 1000–1400 than 2200–0200 (fig. 2.9).[113] Of course, these observations were of foraging parents, which, unlike nonbreeders, cannot choose only the most ideal conditions under which to forage.

Few observations have been made of the diel foraging patterns of nonbreeding Adélies in summer.[7] In the outer pack ice of the Ross Sea well beyond the foraging range of parents and at latitudes higher than those of the Antarctic Peninsula (with mostly continuous light), at-sea observations indicate that lowest proportions of Adélie penguins foraged around midnight and at midday. Lower visibility may have curtailed foraging at midnight, and the usual downward migration of prey when surface light is the most intense may have curtailed foraging at midday.

■ Foraging Cycles and Energetics

The amount of time that an Adélie parent spends away from the nest changes during the nesting period, from egg laying to chick fledging. Some portion of the time away is spent swimming and foraging, but researchers have measured this only during three portions of the annual cycle (late incubation, guard, and crèche periods) and only at locations in and around the northern part of the Antarctic Peninsula. The measurements were made using time-depth recorders. During these three periods in the reproductive cycle at Esperanza Bay, Adélies spent an average ninety-six, thirty-six, and twenty-one hours, respectively, away from their nests in search of food for each trip.[111] For comparison, at other locations, times away during the guard and crèche periods were logged as follows: Admiralty Bay, King George Island, forty-eight and twenty-four hours;[93] Signy Island, South Orkneys, forty-two to fifty-five and thirty-eight to forty-four hours;[62] and Arthur Harbor, Anvers Island, forty and thirty-three hours.[22] At Esperanza (during the three phases of reproduction), of the period away during each trip, 30, 31, and 27 percent of that time was spent foraging underwater (depth greater than 5 meters).[108] At Arthur Harbor, on a typical trip, 52 percent of the time away was spent traveling (swimming shallower than 5 meters), and 32 percent was spent foraging (diving deeper than 5 meters). Of the time foraging (deep dives), 15.4 percent was spent at the bottom of the U-shaped dives (i.e., actually capturing prey).[21]

In sum, although the amount of time spent swimming and diving varies from place to place and year to year, Adélie penguins put a great deal of effort into swimming and diving, especially when foraging to provision chicks. At Arthur Harbor, in the thirty-three hours spent away during the crèche period, a parent spent sixteen hours swimming to and from the feeding area and about ten hours actually diving in search mode or prey capture mode. Recall that at this site, Adélie penguins travel fewer than 15 kilometers to feeding areas; elsewhere they are known to swim much farther during trips to provision chicks. Thus at Arthur Harbor they can devote even more time to swimming and diving.

It is remarkable that an Adélie penguin can spend so much time swimming and diving. How do they do it? Many of the details are poorly known, but several factors determine their swimming capacity. First is the metabolic rate, which in turn dictates how much energy they burn in searching for and capturing prey and how much prey they must capture just to keep searching. Second is the quantity of oxygen they have available to remain underwater on each dive. If their metabolic rate is high, then oxygen is used quickly; if low, then oxygen can be used to extend diving. Physiologists discuss this factor in terms of the aerobic dive limit (ADL),[57] which is the amount of time spent in a dive before the stored oxygen is depleted. To stay under the surface longer, the penguin must burn extra calories and tissues to release oxygen (anaerobic diving). One study, using the doubly-labeled water technique (a measure of how rapidly isotopically labeled oxygen, injected into the blood, is incorporated into body tissues), fixed the Adélie penguin ADL at about forty-eight seconds.[22] This duration was extrapolated using a high estimate of metabolic rate. In that study, the average Adélie foraging dive was seventy-three seconds, so it seems that most dives had to have an anaerobic portion. Another study, using the actual oxygen uptake rate of Adélies swimming in a very long, enclosed dive chamber at the ends of which they could breathe (and gas exchange was measured), fixed the ADL at about 110 seconds.[29] Thus the metabolic rate determined by this method was lower than that determined in the other study. Even in the dive canal, however, 54 percent of dives apparently exceeded the ADL.

It is difficult to understand how the Adélie penguin can undertake prolonged periods of continuous diving, especially if many dives exceed the ADL. Seemingly, they should be incurring a huge oxygen debt. Porpoising for long periods is not difficult to understand. As noted earlier, they swim at the maximum speed that minimizes their cost of transportation, and they do so aerobically (because they take a breath when they have to). To dive for a period beyond the amount of oxygen they have readily stored in their lungs, hemoglobin (blood), and myoglobin (muscles), they must metabolize tissues to free more oxygen. When they do so, lactic acid (product of the metabolic process that releases oxygen from tissues) builds up. The penguin must expel this lactate, or it gets tired and stiff. Therefore anaerobiosis is energetically expensive, so Adélies would not be able to swim anaerobically for long. Foraging range and duration would be strictly limited.

One school of thought surmises that Adélie penguins dispel the lactate loads during the aerobic portions of dives (using oxygen in the process) or they are very efficient at metabolizing lactate during their rest periods.[22] Sometimes between dives, Adélies remain at the surface for about 50 percent of the time that they spent in the previous dive. During this surface period, they can replenish their oxygen stores and metabolize the lactate. This is ecologically expensive, however, because to avoid spending additional time and energy in prey searching, they may need to dive repeatedly, with little rest, to keep track of the prey patch. Therefore they often spend only ten to twenty seconds at the surface, even when making dives two minutes or so in duration. One strategy would be for Adélies (and other diving birds) to forage in bouts (i.e., continuous diving for up to a few hours followed by a rest before the next bout). This would allow them to stay with the prey school more efficiently, but then lactate builds up during the diving bout. In one study of Adélie penguins feeding through tide cracks in fast ice, foraging bouts averaged twenty-five minutes, with about twelve (long) dives per bout.[67] On the other hand, in a study where the penguins were foraging in open pack ice (short dives), they dove continuously for up to eighteen hours.[105]

Another strategy to prevent an oxygen debt from repeated diving would be to lower the metabolic rate. With a lower metabolic rate, less oxygen would be used.[19] A penguin's heart rate decreases by more than 50 percent when it dives (250 down to 107 beats per minute), which could indicate a change in metabolic rate.[26] A lower body temperature would also lower the penguin's metabolic rate.[19,20] However, recovering their body temperature and warming up the near-to-freezing food that they just ingested would be energetically costly.[106] Ultimately, the oxygen-storing capabilities of the penguin's huge muscles and the fact that its brain is small and therefore does not use as much oxygen as a large brain[59] help the penguin manage its store of oxygen. Of course, much more remains to be learned about penguin diving.

Another puzzling aspect of penguin diving is that more than 30 percent of the oxygen used in a dive comes from their lungs and air sacs. The remainder of oxygen is stored in their blood and muscle tissue.[57,76] Descending and ascending as rapidly as they do to the deep depths they reach, with so much air and oxygen in a gaseous state to start, may mean that the free oxygen is used first. In the deeper portion of the dive the oxygen comes from that stored in their very large muscles. Otherwise, they might be prone to the bends (gas bubbles in the blood).[29] Without doubt, they do not have this problem.

■ How Much Prey Must Adélie Penguins Capture?

Adélie penguins consume about 25 grams of krill (one adult *Euphausia superba* every 1.3 seconds) on each dive.[108] This estimate is based on the total number of prey consumed on a foraging trip and the amount of time an Adélie remains in its foraging mode during a trip (i.e., the amount of time at the bottom of U-shaped dives). Similar calculations for African penguins, pursuing fish of about the size that Adélie penguins eat, indicate that they can swallow a fish every five seconds.[114] Because of the variation in the density of prey encountered, however, these probably are maximum rather than average rates of prey capture. Using estimates of penguin metabolic rate (energy needs), the energy content of the prey, the number of dives per foraging trip, and other information, other researchers have estimated that an Adélie needs to catch about ten adult krill per dive (range, perhaps, 4–112).[22,105] On a typical foraging trip when provisioning chicks, if they caught only adult Antarctic krill, they would have to catch several hundred to a thousand or more. Of course, estimating the number that they must capture depends on the physiological rates that are used in the calculations. As we have seen, these rates vary depending on study methods.

Estimates have also been made of the amount of food parents must deliver to raise a chick to fledging. The figures range from 23 kilograms[93] to 33.6 kilograms.[50] A third estimate is midway between these, 29.8 kilograms,[27] and an estimate based purely on modeling is 24.1 kilograms.[85] Parents feed only about half the food they capture to their chicks; the remainder they consume themselves.[12,62] If foraging trips exceed about two days, then less food is delivered to the chicks.

Although these calculations appear to be rather academic (and approximate), there is a practical side to the issue. Without question, an individual Adélie penguin captures a lot of food during the year, with prey availability needing to be especially high when chicks are being provisioned. Moreover, because swimming is so energetically costly and slow, a penguin cannot range as far as a flying bird. Therefore the prey must be in adequate supply close to the colony. When you add up the number of Adélie penguins in a colony, the estimate of prey consumption becomes astronomical. One estimate for all breeding Adélie penguins, with many assumptions built into the calculations, put their annual consumption at 1.5 million metric tons of krill, 115 thousand metric tons of fish, and 3.5 thousand metric tons of squid.[115] Because of such consumption, CCAMLR (the fishery conservation agreement for the Southern Ocean) came into being. Adélie penguins are consummate predators, but so are humans. CCAMLR hopes to protect the penguins (and other species) from competition with humans.

Chapter 3

Breeding Populations

Size and Distribution

The Adélie penguin world population numbers about 2.5 million breeding pairs, which makes this species an abundant seabird. However, it is by no means the most populous seabird species or even the most abundant penguin. More abundant are the macaroni penguin (*Eudyptes chrysolophus*, 11.8 million pairs), chinstrap penguin (7.5 million), and rockhopper penguin (*Eudyptes chrysocome*, 3.7 million).[56]

Adélie penguins nest at sites distributed around the coast of Antarctica, as well as on coasts of islands located at varying distances away from the continent (fig. 3.1). For instance, they nest in the Balleny and South Shetland island groups (110-200 kilometers off the continental shore), the South Orkney Islands (about 500 kilometers), and the South Sandwich Islands (about 1,500 kilometers). The common features of these island and mainland sites are that each is surrounded by pack ice during winter and early spring every year; they are accessible from the sea by walking; they consist of snow-, ice-, and meltwater-free terrain that slopes less than 45°; and there is abundant gravel with which penguins can construct their nests.

Levick, who compiled insights into this species's biology during a long stay at Cape Adare, Victoria Land, noted these characteristics of nesting areas:

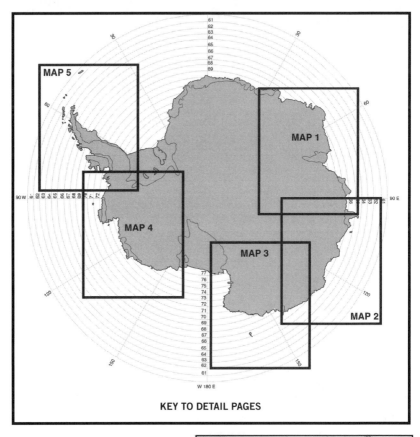

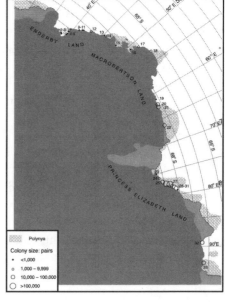

Figure 3.1 (opposite) Maps of Antarctica, offshore islands, and coastal polynyas by sector. Open circles show localities of Adélie penguin colonies, with size of circle proportional to colony size, as of the 1990s. *Data from table 3.1.*

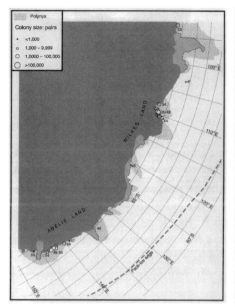
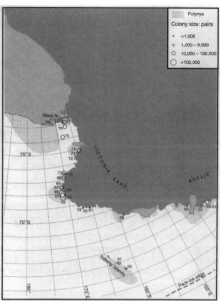
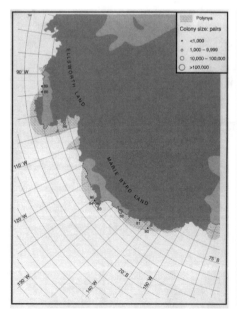
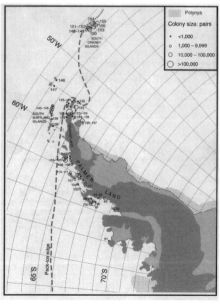

Their requirements are few: they seek no shelter from the terrible Antarctic gales, their rookeries in most cases being in open wind-swept spots. In fact, three of the four rookeries I visited [capes Royds and Adare, Duke of York and Inexpressible islands] were possibly in the three most windy regions of the Antarctic. The reason for this is that only wind-swept places are so kept bare of snow that solid ground and pebbles for making nests are to be found. (p. 7)[22]

Falla[14] noted (p. 76), "It is more significant to record absence of Adélie penguins in the coastal zone of Antarctica than their presence, for there are few rocky capes or islands in the pack-ice zone that do not support rookeries." In other words, sites with the necessary characteristics are not abundant. Therefore it appears that Adélie penguins benefit from snow- and ice-related processes in two ways. The penguins are not just obligate inhabitants of the pack ice, therefore nesting in proximity to it; they also rely on the glaciers, which formed the moraines that deposited nest stones

Illustration 3.1 Almost the entire coastline of Antarctica looks like this: endless ice cliffs, with no place for an Adélie penguin colony. *Drawing by Lucia deLeiris © 1988.*

BREEDING POPULATIONS 73

Illustration 3.2 The type of terrain on which Adélie penguins might establish a colony if pebbles are present. *Drawing by Lucia deLeiris © 1988.*

at most sites on the continent. On islands and along the shores of the Antarctic Peninsula, the erosion of land by ocean waves creates additional nesting stones.

A remarkable example of how Adélie penguins respond to an accumulation of stones is evident in the Weddell Sea region. Emperor penguins, which nest on fast ice along the western edge of the Weddell Sea, occasionally swallow small stones as they forage near the ocean bottom.[37] Perhaps the stones aid in grinding food, much as domestic fowl eat gravel to increase the grinding efficiency of their gizzards. Back at the colony on the sea ice, the penguins regurgitate these stones from time to time. At large emperor penguin colonies, enough stones lie about that Adélie penguins can make nests in which they lay and incubate their own eggs. The remarkable thing is that the nests are situated on the ice; the closest land accessible from the sea is hundreds of kilometers away.

THIS CHAPTER summarizes information on the location and sizes of Adélie penguin breeding sites. The number of birds at one site affects the numbers at nearby sites, depending on the distance between them (and other factors), as explained later in this chapter. First, it is necessary to understand what defines an Adélie penguin colony. Eventually we will discuss the factors that affect colony location. In chapter 9, the discussion turns to the appearance and disappearance of colonies caused by changes in the environment.

■ What Is a Penguin Colony?

A colony is an assemblage of nesting Adélie penguins. More specifically, the term *colony* refers to all Adélie penguins breeding within a 5-nautical-mile (8–kilometer) radius who are strongly related demographically. My reasons for this definition are explained later in more detail. Such assemblages have been called rookeries by many researchers,[13,29,35,44] including me.[3] The term *rookery* is left over from the heroic period of Antarctic exploration,[22,23] when Antarctic fur seal "rookeries" were exploited commercially. However, Pauline Reilly,[32] who has studied the little blue penguin in Australia for many years during modern times, advocates using the term *colony* to name the geographic entity under discussion here. Penguins are seabirds, and every other seabird species that breeds colonially, as does a penguin, does so in what are called colonies; in other words, a colony is what a colonial breeder nests in. Why should penguins be described differently? To avoid confusion and to encourage comparison of natural history patterns between all seabird species, I use the term *colony* and encourage others to follow Pauline Reilly's lead. Use of the term *colony* in the sense proposed here for the Adélie penguin is not without precedent.[5,12,42]

Within a colony, penguins nest in groups called subcolonies. Within the groups, territories are contiguous. That is, the outer edge of one territory abuts the outer edge of at least one other territory. If it stretches full length, a penguin sitting on its nest can catch and lock its beak with that of its neighbor, also stretching full length from its nest. A territory within a breeding group might abut as many as six other territories. Formerly, in the case of penguins but not other seabirds, these groups of contiguous territories were called colonies.[3,29,35,57] To prevent confusion, I refer to these groups within a colony as *subcolonies* or breeding groups (fig. 3.2).

Confusion over what constitutes a colony or subcolony has been widespread, as exemplified in the useful summaries of the distribution of penguin populations published by the Scientific Committee for Antarctic Research (SCAR).[33,34,55] In these summaries, Eric Woehler avoided reinterpreting the way in which *colony* and *subcolony* were defined by the individual authors of some 130 publications he reviewed. As a result, some

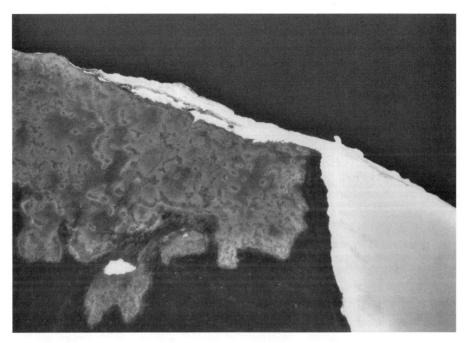

Figure 3.2 Aerial image of an Adélie penguin colony showing clusters of contiguous territories, which are called subcolonies. Each black dot is a penguin.

localities are said to comprise one colony (e.g., Beaufort Island) whereas others of similar size comprise eighty-three colonies (e.g., Peterson Island; table 3.1). It is obvious, in this comparative example, that the author who originally reported the counts of breeding pairs for Beaufort Island was using *colony*, in the new sense, to denote the total of the breeding groups (or subcolonies), whereas the author reporting counts for Peterson Island was using *colony* in the old sense (i.e., colonies referred to as subcolonies).

My model for defining the term *colony* in the case of the Adélie penguin is the one existing at Beaufort Island (76°56'S, 167°03'E), in the southern Ross Sea. On this island, there is only one Adélie penguin breeding assemblage, and the next closest one is 31 kilometers away (fig. 3.3A). In other words, the colony clearly is isolated from any others. Studies of banded penguins have shown that at this distance, fewer than 0.01 percent of penguins from the next closest colony might be expected to immigrate to Beaufort Island under normal circumstances. That is, raised as a chick in a nest in one colony, most birds will return to nest there as

Table 3.1. Catalog of Adélie penguin colonies.

SCAR No.	Map No.	Colony Name	Location	Number of Localities	Number of Pairs	Polynya Adjacent
Dronning Maud Land						
3	1	Schirmacher	70°45′S, 11°40′E	1	3	
Enderby Land						
5	2	Meholmen	68°58′S, 39°32′E	1	1	
6a	3	Kujira Point	69°36′S, 39°16′E	1	30	
6b	4	Torinosu Cove	69°29′S, 39°34′E	1	135	
6c	5	Nokkelhomane Is	69°24′S, 39°29′E	1	100	
6d	6	Hukuro Cove	69°12′S, 39°39′E	3	445	
6e	7	Rumpa Is	69°08′S, 39°26′E	1	1,250	
6f	8	Benten Is	69°01′S, 39°13′E	3	155	
7a	9	Cape Omega	68°34′S, 40°59′E	1	250	Syowa
7b	10	Akarui Point	68°29′S, 41°23′E	1	170	Syowa
8	11	Tenmondai Rock	68°25′S, 41°41′E	1	19	Syowa
8a	12	Cape Hinode	68°07′S, 42°38′E	1	150	Syowa
9	13	Alasheyev Bight	67°30′S, 45°40′E	1	1,000	Syowa
9a–c	14	Myall/McMahon	67°39′S, 45°50′E	3	2,316	Syowa
MacRobertson Land						
10	15	Casey Bay	67°30′S, 48°00′E	1	?	Casey Bay
10a	16	Mt Riiser–Larsen	66°47′S, 50°41′E	1	877	Casey Bay
11	17	Mt Biscoe	66°13′S, 51°22′E	2	5,000	Enderby Land
12–14	18	Proclamation Is	65°51′S, 53°41′E	3	5,000	Enderby Land
15	19	Kidson Is	67°12′S, 61°11′E	1	1,000	Enderby Land
16a	20	Rookery Is	67°37′S, 62°32′E	?	44,800	Taylor Glacier
16b	21	Mawson Area	67°35′S, 62°45′E	?	37,695	Taylor Glacier
17–18	22	Scullin/Murray Monolith	67°47′S, 66°48′E	2	69,500	Cape Darnely
Princess Elizabeth Land						
20	23	Bolingen/Lichen Is	69°24′S, 75°38′E	2	465	Prydz Bay
	24	Steinnes Is	69°22′S, 76°34′E	2	1,176	Prydz Bay
	25	Brattstrand Bluff/Svenner Is	69°07′S, 76°55′E	8	41,389	Prydz Bay
21	26	Rauer Is	68°51′S, 77°50′E	11	103,916	Prydz Bay

Table 3.1. Continued

SCAR No.	Map No.	Colony Name	Location	Number of Localities	Number of Pairs	Polynya Adjacent
22	27	Vestfold South	68°35′S, 77°55′E	13	132,049	Prydz Bay
	28	Vestfold North	68°28′S, 78°05′E	4	37,510	Prydz Bay
	29	Tyrne/Long Peninsula Is	68°23′S, 78°23′E	6	24,229	Prydz Bay
	30	Wyatt Earp Is	68°22′S, 78°32′E	3	2,804	Prydz Bay
	31	McCallie Rocks		?	?	Prydz Bay
23	32	Gaussberg	66°48′S, 89°11′E	1	7,500	
24	33	Haswell Is	66°31′S, 93°00′E	4	11,300	Shackleton

Wilkes Land

SCAR No.	Map No.	Colony Name	Location	Number of Localities	Number of Pairs	Polynya Adjacent
25	34	Davis Is	66°40′S, 108°25′E	1	8,730	Cape Poinsett
39	35	Peterson Is	66°28′S, 110°30′E	1	20,453	Cape Poinsett
37–38	36	Holl/O'Connor	66°25′S, 110°26′E	2	16,623	Cape Poinsett
36	37	Odbert Is	66°22′S, 110°33′E	1	10,689	Cape Poinsett
34–35	38	Hollin/Midgley	66°20′S, 110°24′E	2	10,237	Cape Poinsett
32–33	39	Shirley/Beall	66°18′S, 110°30′E	2	12,861	Cape Poinsett
30–31	40	Blakeney/Whitney	66°14′S, 110°34′E	2	9,407	Cape Poinsett
28–29	41	Berkley/Cameron	66°13′S, 110°38′E	2	6,488	Cape Poinsett
27	42	Chappel Is	66°11′S, 110°26′E	1	5,780	Cape Poinsett
26	43	Nelly Is	66°14′S, 110°11′E	1	554	Cape Poinsett
40	44	Balaena Is	66°01′S, 111°06′E	1	?	Cape Poinsett
41	45	Chick Is	66°44′S, 121°00′E	?	273	Dalton

Adélie Land

SCAR No.	Map No.	Colony Name	Location	Number of Localities	Number of Pairs	Polynya Adjacent
42	46	Lewis Is	66°06′S, 134°22′E	1	1,200	Dibble
47	47	Rochers	66°33′S, 139°10′E	2	420	
43–44	48	Point Géologie/Bienvenue	66°42′S, 140°15′E	?	34,000	Mertz
45	49	Cape Jules	66°44′S, 140°55′E	1	44,300	Mertz
46	50	Port Martin	66°49′S, 141°24′E	8	55,860	Mertz
48–49	51	Cape Denison/MacKellar Is	66°59′S, 142°40′E	2	32,158	Mertz
50	52	Way Archipelago	66°53′S, 143°40′E	?	23,084	Mertz
50a	53	Cape Pigeon Rocks	66°59′S, 143°47′E	9	10,600	Mertz
51	54	Watt Bay	67°02′S, 144°00′E	2	1,850	Mertz
50b	55	Penguin Point	67°39′S, 146°12′E	1	21	Mertz
52	56	Aviation Is	69°16′S, 158°47′E	3	1,093	Terranova Is

BREEDING POPULATIONS

Table 3.1. Continued

SCAR No.	Map No.	Colony Name	Location	Number of Localities	Number of Pairs	Polynya Adjacent
Balleny Islands						
53–55	57	Balleny Is	66°55′S, 163°20′E	5	6,830	Balleny
56	58	Sturge Is	67°28′S, 164°38′E	1	10	Balleny
Victoria Land						
57	59	Nella Is	70°37′S, 166°04′E	1	181	Ross Passage
58	60	Unger Is	70°41′S, 166°55′E	1	166	Ross Passage
59	61	Sentry Rocks	70°45′S, 167°24′E	1	119	Ross Passage
60	62	Duke of York Is	71°38′S, 170°04′E	1	2,307	Ross Passage
61	63	Cape Adare	71°18′S, 170°09′E	1	169,200	Ross Passage
62	64	Downshire Cliffs	71°37′S, 170°36′E	1	12,492	Ross Passage
63–64	65	Possession Is	71°54′S, 171°08′E	2	162,070	Ross Passage
65	66	Cape Hallett	72°19′S, 170°16′E	1	43,942	Ross Passage
66	67	Cotter Cliffs	72°28′S, 170°28′E	1	27,764	Ross Passage
67	68	Cape Wheatstone	72°37′S, 170°13′E	1	1,733	Ross Passage
68–69	69	Cape Phillips/ Mandible Cirque	73°06′S, 169°26′E	2	22,610	
70	70	Cape Jones	73°17′S, 169°10′E	1	101	
71	71	Coulman Is	73°28′S, 169°45′E	4	21,887	
72	72	Wood Bay	74°19′S, 165°04′E	1	1,995	
73	73	Terra Nova Bay	74°45′S, 165°06′E	1	7,899	Terra Nova
74	74	Inexpressible Is	74°53′S, 163°45′E	1	20,029	Terra Nova
75	75	Franklin Is	76°05′S, 168°19′E	2	55,600	Ross Sea
76	76	Beaufort Is	76°56′S, 167°03′E	1	37,668	Ross Sea
77	77	Cape Bird	77°13′S, 166°28′E	3	35,732	Ross Sea
78–79	78	Cape Barne/Royds	77°34′S, 166°11′E	2	4,096	Ross Sea
80	79	Cape Crozier	77°31′S, 169°23′E	2	118,120	Ross Sea
Marie Byrd Land						
81	80	Cruzen Is	74°47′S, 140°42′W	1	135	Ruppert
82	81	Cape Burks	74°45′S, 136°50′W	1	97	Ruppert
83	82	Shepard Is	74°25′S, 132°30′W	2	40,000	Ruppert
84	83	Lovill Bluff	73°22′S, 126°54′W	2	61,533	Amundsen
85	84	Maher Is	72°58′S, 126°22′W	1	4,722	Amundsen
86	85	Lauff Is	73°03′S, 126°08′W	1	1,167	Amundsen

Table 3.1. Continued

SCAR No.	Map No.	Colony Name	Location	Number of Localities	Number of Pairs	Polynya Adjacent
Ellsworth Land						
86b	86	Lindsey Is	73°37'S, 103°18'W	8	5,000	
86c	87	Thurston Is	72°06'S, 99°00'W	1	?	
86d	88	Dustin Is	72°34'S, 94°50'W	1	?	
86e	89	McNamara Is	72°34'S, 93°12'W	1	?	
87	90	Peter Is	68°47'S, 90°35'W	1	69	
Palmer Land						
88	91	Charcot Is	69°45'S, 75°15'W	1	50	
89a	92	Rhyolite Is	69°40'S, 68°35'W	1	35	
89b	93	Red Rock Ridge	68°18'S, 67°11'W	1	1,200	Marguerite Bay
90	94	Lagotellerie Is	67°53'S, 67°23'W	1	1,700	Marguerite Bay
91	95	Pourquoi Pas Is	67°44'S, 67°28'W	1	700	Marguerite Bay
92	96	Emperor Is	67°52'S, 68°43'W	1	700	Marguerite Bay
93	97	Avian Is	67°46'S, 68°54'W	1	35,600	Marguerite Bay
94	98	Adelaide Is	67°45'S, 69°00'W		?	Marguerite Bay
95	99	Ginger Is	67°45'S, 68°42'W	1	3,000	Marguerite Bay
96–97	100	Cone/Chatos Is	67°40'S, 69°10'W	2	3,100	Marguerite Bay
98–100	101	Holdfast Point and Islands	66°48'S, 66°35'W	3	3,725	
101	102	Barcroft Is	66°27'S, 66°10'W	1	1,600	
102	103	Darbel Is	66°23'S, 65°58'W	1	650	
103	104	Lavoisier Is	66°12'S, 66°44'W	1	150	
104	105	Cape Evensen	66°09'S, 65°44'W	1	1,100	
105	106	Fish Is	66°01'S, 65°21'W	1	4,000	
106	107	Armstrong Reef	65°54'S, 66°18'W	1	12,800	
107	108	Kim Is	65°41'S, 65°20'W	1	1,300	Palmer Deep
108	109	Vieuque Is	65°39'S, 65°14'W	1	1,000	Palmer Deep
109–110	110	S Pitt/Fitzkin Is	65°30'S, 65°36'W	2	300	Palmer Deep
111–112	111	N Pitt/Trundle Is	65°24'S, 65°20'W	2	4,550	Palmer Deep
113	112	Berthelot Is	65°20'S, 64°08'W	1	1,300	Palmer Deep
114	113	Yalour Is	65°15'S, 64°11'W	1	8,000	Palmer Deep
115	114	Petermann	65°11'S, 64°10'W	1	1,080	Palmer Deep
116–117	115	Booth Is/Port Charcot	65°04'S, 64°00'W	2	1,269	Palmer Deep

Table 3.1. Continued

SCAR No.	Map No.	Colony Name	Location	Number of Localities	Number of Pairs	Polynya Adjacent
118	116	Biscoe Point	64°49′S, 63°49′W	1	3,500	Palmer Deep
119–124	117	Arthur Harbor	64°46′S, 64°04′W	5	14,170	Palmer Deep
125	118	Dream Is	64°44′S, 64°15′W	1	11,263	Palmer Deep
126	119	Joubin Is	64°47′S, 64°27′W	4	1,251	Palmer Deep
127	120	Gerlache Is	64°36′S, 64°15′W	1	171	Palmer Deep

Graham Land

SCAR No.	Map No.	Colony Name	Location	Number of Localities	Number of Pairs
128	121	Duroch Is	63°18′S, 57°50′W	1	800
129	122	Gourdin Is	63°12′S, 57°15′W	1	300
130a	123	Hope Bay	63°23′S, 57°00′W	1	123,850
130b	124	Tabarin Pen	63°32′S, 56°55′W	1	20,000
131	125	Jonassen Is	63°33′S, 56°40′W	1	?
132	126	Bransfield Is	63°11′S, 56°35′W	1	100
133	127	d'Urville Is	63°05′S, 56°20′W	1	70
134	128	Wideopen Is	63°00′S, 55°51′W	1	100
135	129	Patella Is	63°08′S, 55°31′W	1	>1,000
136	130	Joinville Is	63°18′S, 56°29′W	2	35,000
137	131	Etna Is	63°06′S, 55°10′W	1	25
138	132	Danger Is	63°24′S, 54°38′W	1	15,000
139	133	Paulet Is	63°35′S, 55°46′W	1	60,000
140	134	Vortex Is	63°44′S, 57°38′W	1	300
142	135	Devil Is	63°48′S, 57°17′W	1	10,320
141	136	Cockburn Is	64°12′S, 56°50′W	1	?
143	137	Seymour Is	64°18′S, 56°45′W	2	21,954

South Shetland Islands

SCAR No.	Map No.	Colony Name	Location	Number of Localities	Number of Pairs
144	138	Hannah Point	62°39′S, 60°37′W	1	2
145	139	Nelson	62°18′S, 59°14′W	1	?
146	140	Llano Point	62°11′S, 58°27′W	1	6,100
147	141	Ardley Is	62°13′S, 58°57′W	1	1,192
148	142	Stranger Point	62°16′S, 58°37′W	1	14,554
149	143	Point Thomas	62°10′S, 58°27′W	1	8,645
150	144	Lions Rump	62°08′S, 58°08′W	1	12,345
151–153	145	Penguin/Turret/3 Sisters	62°05′S, 57°56′W	3	11,234
154	146	Gibbs Is	61°30′S, 55°29′W	1	2
155	147	Clarence Is	61°19′S, 54°06′W	1	119

Table 3.1. Continued

SCAR No.	Map No.	Colony Name	Location	Number of Localities	Number of Pairs	Polynya Adjacent
South Orkney Islands						
156	148	Gosling Is	60°38'S, 45°55'W	1	5,700	
157–159	149	Cape Hansen/Stene/Shingle	60°40'S, 45°36'W	3	4,500	
160	150	Amphibolite Point	60°41'S, 45°20'W	1	4,000	
161–162	151	North Point/Spindrift	60°40'S, 45°38'W	2	17,525	
163	152	Gourlay Pen	60°44'S, 45°35'W	3	37,200	
164–166	153	Powell/Michelsen/Christoffersen	60°44'S, 45°02'W	4	16,750	
167	154	Watson Pen	60°40'S, 44°31'W	1	462	
168–170	155	Ferrier/Graptolite/Fitchie	60°43'S, 44°26'W	5	94,000	
171	156	Port Martin	60°46'S, 44°42'W	4	27,958	
South Sandwich Islands						
172	157	Thule Is	59°28'S, 27°15'W	1	10,000	
173	158	Bellingshausen Is	59°25'S, 27°03'W	2	10,000	
174	159	Montagu Is	58°25'S, 26°20'W	1	200	
175	160	Saunders Is	57°45'S, 26°27'W	1	50,000	
176	161	Candlemas Is	57°05'S, 26°40'W	1	1,200	

Breeding localities are combined into colonies (if necessary) according to the rules explained in the text. For colonies that represent several localities, latitude and longitude are a compromise indicating the approximate middle of the group. Colonies are secondarily grouped by metapopulation cluster according to rules also explained in the text (clusters include the colonies listed within brackets). Colony numbers refer to numbers in figure 3.1; SCAR numbers refer to the colonies listed in colony catalogs produced by the Scientific Committee for Antarctic Research.[55] Colony count data are from several sources.[19,33,34,53,55] If known, the polynya immediately adjacent to at least one colony in a cluster is indicated. Polynya location data are from several sources.[21,24–27,39,41,43,46,48,58]

adults and few will emigrate to stake out a territory and breed at Beaufort Island. There are variations on this colony theme—the clearly isolated, breeding assemblage—having to do with the proximity of neighboring colonies and the expected degree of interchange of penguins between them. These are reviewed later in this chapter. Among Adélie penguin

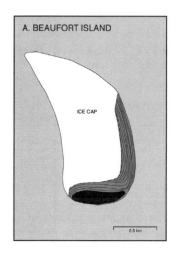
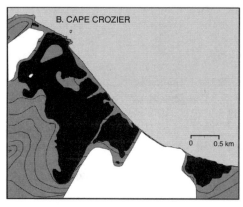
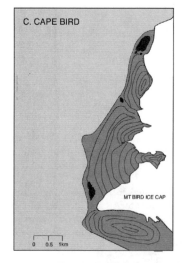
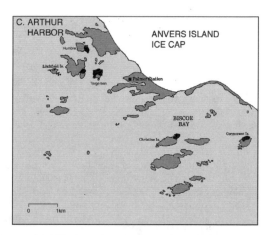

Figure 3.3 Variations on the layout of Adélie penguin colonies. A) Beaufort Island is a single and well-isolated assemblage of nests. B) Cape Crozier East and West are just 2 kilometers apart and served by separate landing beaches, but there is substantial interchange of adults hatched in one but breeding in the other. C) At Cape Bird, three distinct assemblages are served by the same landing beach (on a geographic scale similar to that of Cape Crozier and Cape Adare, two of the world's largest Adélie penguin colonies). D) At Arthur Harbor, Anvers Island, groups of subcolonies nest on separate islands within an area equivalent to that of Cape Crozier or Cape Adare.

breeding assemblages, there are many situations similar to that of Beaufort Island; I chose it and the other variants described here as examples because I am personally familiar with them.

The first variant on the Beaufort model is exemplified by Cape Crozier on Ross Island, also in the southern Ross Sea. At Cape Crozier (77°31'S, 169°23'E), one large assemblage of breeding penguins (about 120,000 pairs) is separated from another assemblage (about 30,000 pairs) by a 2–kilometer stretch of ice and rock cliffs too steep for an Adélie penguin to climb (fig. 3.3B). Penguins do not travel from one assemblage to the other except by water because the terrain between is too rugged for them to walk. Studies of banded birds have shown that Crozier East and Crozier West are close enough that as many as 0.25 percent of chicks reared in Crozier West would be expected to emigrate to Crozier East.[1] That percentage of intermixing is large for seabirds, which as a rule do not tend to disperse from natal colonies. Because of the strong demographic relatedness of these two assemblages, I consider them to be one colony.

The next variant of the colony theme is the one at Cape Bird, also on Ross Island but 60 kilometers from Cape Crozier (and 31 kilometers from Beaufort Island). At Cape Bird (77°33'S, 166°28'E), there are three disjunct breeding assemblages about 1 kilometer apart (fig. 3.3C). Other people have referred to these as the North, Middle, and South colonies of Cape Bird (or the North, Middle, and South rookeries),[54] but all three breeding assemblages are served by a single landing beach, which stretches the length of the cape. I have seen penguins from each area assemble at the beach, intermix, and then depart for the sea in one flock. Flocks of intermixed individuals arrive back after feeding, then separate, either while swimming just off the beach or after landing. This same sort of intermixing and separating of foraging flocks occurs at Cape Crozier East and West, except that at Crozier if an East Crozier bird lands at West Crozier, it has to go back into the water to reach East Crozier. At Cape Bird it could walk to whatever assemblage contained its nest. There is likely to be a great deal of interchange between these assemblages at Cape Bird. That is, a penguin raised within one assemblage (say, Cape Bird North) might nest in another (say, Cape Bird Middle). No information is available to substantiate this idea at Cape Bird, but because the geographic scale is equivalent to that of Cape Crozier, and I have seen this intermixing of foraging individuals there, I combine the three breeding groups at Cape Bird into one colony. (Current research using banded individuals, yet to be published, confirms the intermixing).

The third variant of the model may be controversial among penguin researchers because it represents a departure from current practice. I am proposing to combine into one colony assemblages that exist on separate but closely clumped islands. As an example, I use the collection of breeding assemblages on separate small islands in Arthur Harbor, Anvers Island (64°46'S, 64°04'W). My reasoning is as follows. The largest population of Adélie penguins nesting at a single, uninterrupted geographic location is that at Cape Adare, Victoria Land (71°18'S, 170°09'E; table 3.1). During the late 1980s, the Cape Adare breeding population was estimated to be 282,307 pairs.[44] This colony is roughly an equilateral triangle with a base of 2.3 kilometers and a height of 1.5 kilometers.[31] Therefore the area of the Cape Adare colony is about 1.7 square kilometers, which easily spans any two of the three Cape Bird breeding localities described earlier. Cape Crozier West spans a similar area (its subcolonies are more closely spaced than those at Cape Adare). Crozier West and Crozier East extend about 4 kilometers in one dimension and about 1 kilometer in the other (at their broadest), with a total area of about 2.5 square kilometers. As noted earlier, the two Crozier assemblages make up one colony, demonstrated demographically.

In Arthur Harbor and immediately to the southeast (Biscoe Bay), there are five islands on which Adélie penguins nest and several other islands where none nest (about seventeen islands total). The five islands are spread in a line roughly parallel to the shore of the much larger Anvers Island over a span of about 6 kilometers (centered at about 64°46'S, 64°03'W). Considering this length and the broadest width among the colonies (Litchfield Island to Torgersen Island), the total area is 2-3 square kilometers, roughly equivalent to Cape Crozier (fig. 3.3D). On each of these five islands are one to five breeding groups (or subcolonies). A banding study begun by William Fraser and me and continued by Fraser has shown appreciable interchange between these islands, similar to that evident between Crozier East and Crozier West. Therefore I consider the assemblages on these five islands to make up one colony, even though they have separate landing beaches (as at Cape Crozier East and West). Other researchers consider these assemblages in Arthur Harbor to be separate colonies.[18,28] The Arthur Harbor colony (my definition) is separated from other Adélie penguin colonies by 8 kilometers or more, a stretch that includes islands on which the species does not nest.

In many cases, these examples duplicate what has been done in a catalog of penguin colonies, although the author[55] did not explain the rules used in that case. For example, the catalog specifies just one breeding "location" at capes Crozier and Bird made up of two and three colonies, respectively. In the case of Arthur Harbor and Biscoe Bay, both alternatives are presented (i.e., a total for all five islands and the islands separately). Herein, I have composed a series of rules that I follow in presenting a catalog of colonies. As an example of how I use these rules to identify Adélie penguin colonies, consider the breeding assemblages among the Windmill Islands, Wilkes Land (fig. 3.4). In this region, which stretches about 120 kilometers along the coast of East Antarctica (centered at about 63°20′S, 110°30′E), eighteen breeding locations have been identified.[57] After combining locations on different shores of the same small island or on shores of a group of small islands, in all cases about 5 kilometers or less apart, I come up with ten colonies. Because of the confusion described earlier in the use of the term *colony*, Woehler[55] tallies 568 "colonies" among the eighteen breeding locations.

A scan of colony locations described in Woehler,[55] considered from a large geographic scale (hundreds of square kilometers), shows that combining breeding localities into colonies on the basis of my rules entails combining isolated groups of breeding localities. From a large-scale, regional perspective, the groups I combined on a semidemographic basis are natural; in many cases there are long stretches of colony-free Antarctic coastline between them (see fig. 3.1).

■ Adélie Penguin Colonies and Their Distribution in Antarctica

Distributed on Antarctica and its offshore islands are 161 nesting colonies of the Adélie penguin, totaling about 2,445,000 nesting pairs as of the mid-1990s. These colonies, defined according to the rules just outlined, are summarized in table 3.1. As a result of my rules for defining a colony, this list is about fifteen colonies shorter than the SCAR list.[33,34,55] The SCAR list includes a few sites for which almost no information is known other than a report that Adélie penguins nest there. These sites are not included in table 3.1. Most Adélie penguin colonies are known, but there are probably undiscovered colonies also; such colonies probably are not large. Among the known colonies, the size of ten shown in table 3.1 has not been estimated, and a few others have not been counted for decades. Otherwise, most of the counts were made during the last ten years, thanks to an effort by SCAR, CCAMLR, and numerous other re-

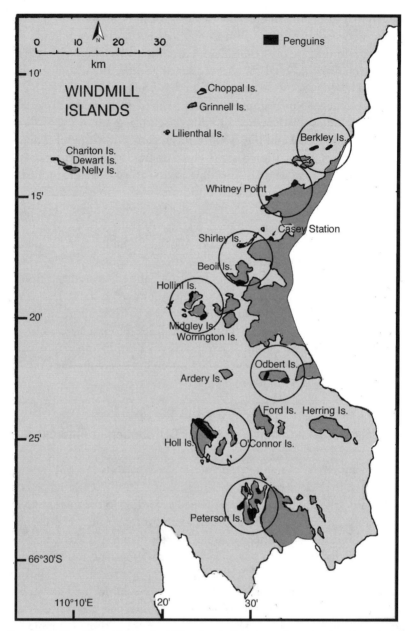

Figure 3.4 The Windmill Islands, Wilkes Land, East Antarctica, showing the colonies identified by Woehler et al.[57] Adjacent colonies combined into what I consider demographic units (i.e., colonies as defined in the text). *The map is drawn after that contained in Woelher et al.*[57]

searchers. The current total should not be considered absolute. Colonies fluctuate in size, with the mid-1990s world total a few hundred thousand pairs fewer than the total in the late 1980s. Also, various researchers use different methods to estimate numbers, and in some cases the estimate is very rough.[55] Therefore an estimate of Adélie penguin pairs today would vary between 2.4 and 3.2 million worldwide.

The smallest colonies consist of only a few pairs of Adélie penguins. In many of these cases, a few Adélies are sprinkled among larger numbers of chinstrap or gentoo penguins. Among the colonies for which estimates are known, nineteen colonies (11.8 percent) have fewer than 100 pairs, and thirty (18.6 percent) have 100 to 1,000 pairs. The lower limit of Adélie penguin colony size probably is governed by proximity to a larger colony and by a minimal number needed to withstand predation from skuas. Many of the colonies smaller than 100 pairs actually consist of Adélie penguins mixed in larger colonies of chinstrap penguins in the Antarctic Peninsula and Scotia Sea region. I discuss factors that may affect the minimal size of colonies in chapter 8.

Just as there are few very small colonies, there are even fewer very large colonies. Only six colonies (3.7 percent) surpass 100,000 pairs. On the basis of the most recent information, the largest colony is that at Ridgley Beach on Cape Adare (71°18′S, 170°09′E), in the Ross Sea, estimated in the early 1990s to total 169,200 pairs.[33] This figure is down from 282,307 pairs counted in the mid-1980s. Not far from Cape Adare is the second largest colony, that of the Possession Islands (including Sven Foyn Island; 71°54′S, 171°08′E), totaling 162,070 pairs in 1990 (and 187,500 a few years earlier; see description by Ross, ch. 1). Also in the Ross Sea is Cape Crozier, another huge colony numbering 118,120 pairs in 1991 (177,083 in the mid-1980s). The world's third largest Adélie penguin colony is known as Vestfold Hills South. It is on the shores of Prydz Bay (68°35′S, 77°55′E), estimated in the late 1980s to total 196,592 pairs.[55] This colony is dispersed over several closely lying capes and offshore islands. Not far from Vestfold Hills is the Rauer Island complex of colonies (68°51′S, 77°50′E), totaling 103,916 pairs in 1981. The remaining large colony, fourth largest in the world and the only very large one in the Antarctic Peninsula region (and, for that matter, in the Western Hemisphere), is that at Hope Bay (63°23′S, 57°00′W, where Sladen[35] conducted his pioneering work; see ch. 1). In 1985, when the last published estimate of its size was made, it totaled 123,850 pairs. Factors that limit the maximum size of an Adélie penguin colony probably include food supply and the degree to which penguins interfere with one another to exploit it; this topic is discussed further in chapter 9.

That leaves us with one hundred seven colonies (66.5 percent) that are between 1,000 and 100,000 pairs. A colony of 1,000 pairs is about the size of the largest subcolonies within the larger colonies. Somewhere in this range must be the ideal in colony size for this species. However, because each colony is a response to the conditions of the region, there may be no ideal size.

Adélie penguin colonies are not distributed evenly in Antarctica. Very few colonies exist in the vast stretches of ice cliffs that border the Weddell Sea. In contrast is the other great sea of Antarctica, the Ross Sea, on the opposite side of Antarctica. Along the Victoria Land coast, the western border of the Ross Sea, we find more than 744,000 nesting pairs of this species, or about 30 percent of the world's breeding population. In the Ross Sea, all these Adélie penguins are found in just twenty-one colonies, or just 13 percent of all colonies. Another area important to the Adélie penguin is Prydz Bay, located along the southern edge of the Indian Ocean midway between the Weddell Sea and the Ross Sea. There, in just eleven colonies (6.8 percent of the world total of colonies), are an additional 395,000 nesting pairs, or about 16 percent of the world breeding population. These two areas combined, with only about 20 percent of the colonies, hold about half the world's population of this species. Five of the six colonies greater than 100,000 pairs are found in these two areas. Together these two areas contribute less than 5 percent of the Antarctic coastline. Why are the Ross Sea and Prydz Bay so special to this species? I will attempt to answer this question after considering other ways in which Adélie penguin nesting populations are grouped.

■ Geographic Structure

As shown in figure 3.1, we find that one or two very large colonies usually occur in conjunction with several smaller colonies nearby. These clumps or clusters, as I observe them, are listed in table 3.1. Thus Adélie penguin colonies are distributed in a recurring pattern, which suggests that the distribution of nesting Adélie penguins is geographically structured. By *geographically structured* I mean the way in which colonies are distributed in terms of their size and distance apart. Studies of banded birds in the southern Ross Sea and the cluster of colonies that includes the large Cape Crozier colony indicate intercolony movement by individuals within that cluster.[3] The southern Ross Sea cluster spans almost 1° of latitude (76°56' to 77°35'S) and more than 3° of longitude (166°09' to 169°23'E). This translates to an area about 70 kilometers on a side, or 4,900 square kilometers. The other clusters listed in table 3.1 are of a similar spatial scale.

A statistical analysis of geographic structuring among Adélie penguin colonies showed that only smaller colonies exist within a radius of 150–200 kilometers of a large colony.[5] The colonies considered in that study were in the Ross Sea and along the Antarctic Peninsula. However, table 3.1 shows that the pattern occurs everywhere around the Antarctic continent. The pattern of clustering in which only small colonies are adjacent to large ones is called negative geographic structuring.

The reasons for this negative geographic structuring of Adélie penguin colonies revolve around two factors: natal philopatry, or the tendency of Adélie penguins to breed close to their birthplace, and depletion of prey resources or access to prey resources to a degree proportional to the number of foraging birds in a given area.

The first factor, natal philopatry has a positive effect on the clumping of colonies. If a colony is large because local conditions (e.g., food, climate, and availability of stones and mates) are especially favorable for penguins, then the tendency to emigrate is low. Why would a recruit go far from home to find a nesting area? One reason to emigrate would be that nesting space has become limited because the colony has grown or some catastrophe (such as increased sea level or the blocking of colony access by a grounded iceberg) has occurred. The penguins that do emigrate will not go very far as long as local conditions remain favorable. As long as suitable nesting habitat is available, new colonies will be founded nearby.

The second factor, prey availability, has a negative effect on the clumping of colonies. Because food supplies are not limitless, a colony will grow until the foraging birds in the region begin to deplete food or foraging birds from adjacent colonies interfere with one another as they attempt to capture prey. In other words, even in the presence of limitless nesting space, the food supply within foraging range of the colony can support only a limited number of penguins (and other top-trophic-level predators). Obtaining adequate food becomes more difficult as penguin numbers in the common foraging area grow or as the foraging area expands to include more individuals.

In a study of foraging areas among adjacent colonies in the southern

Ross Sea cluster, it was found that penguins from the small colonies foraged in a common area. Therefore if one colony is large because it has the most favorable conditions, adjacent colonies within its foraging range can grow only to a level that can be accommodated by the food supply. Generally, then, the other colonies in the cluster are smaller. Of course, this discussion is somewhat theoretical, but there is substantial supporting evidence for geographic structuring for Adélies and other seabirds.[5,10,16]

Earlier it was noted that a given colony's influence on the size of adjacent colonies was detectable within a radius of 150–200 kilometers of the colony. How can that be if the maximum foraging range of an Adélie penguin is 40–50 kilometers while provisioning chicks? The answer is that more than just parents are foraging in proximity to colonies. As noted in chapter 6, nonbreeders (subadults) who visit the colony briefly during the breeding season and juveniles who originated from a given colony but who do not visit all make up the "halo" of foraging birds surrounding an Adélie penguin breeding colony.[6,9] Their total numbers contribute to the density of foraging birds in the vicinity of a colony.

■ Factors Affecting the Distribution of Breeding Colonies

As mentioned earlier, Adélie penguins are obligate inhabitants of the pack ice that surrounds the Antarctic continent. This characteristic explains why Adélie penguins do not nest on islands in the northern portion of the Southern Ocean (i.e., the ocean south of the Polar Front; see ch. 2), islands such as South Georgia, Heard, Kerguelan, Crozet, and Prince Edward. At least since the Last Glacial Maximum (LGM), pack ice has not occurred anywhere near these "antarctic" islands. Therefore it would be a long commute (hundreds of kilometers) for an Adélie penguin to travel from a nesting colony on one of these islands to the nearest sea ice. Where sea ice is a regular feature of the environment, however, Adélie penguins proliferate.

A look at a map of the Antarctic coastline reveals the location of Adélie penguin colonies, colony clusters, and great stretches where no colonies exist (fig. 3.1). By far, the most important factor affecting this pattern is the rarity of ice- and snow-free terrain that is easily accessible by a penguin. Antarctic birds that fly can breed hundreds of kilometers inland on ice- and snow-free mountaintops. This is not possible for penguins. Except in the Antarctic Peninsula region, where most of the ice- and snow-free terrain of Antarctica exists (less than 2 percent of Antarctica is ice-free), almost all of the coast of continental Antarctica ends in ice cliffs, tens to hundreds of meters high, where continental and alpine glaciers meet the sea. No Adélie penguin breeding habitat exists in such places. Where ice-free terrain does exist, in many cases it meets the sea as a cliff in its own right, and this is especially true in the mountainous Antarctic Peninsula region. Thus the few places in the Antarctic where ice-free terrain meets the sea and there is a beach are important to Adélie penguins.

LAND ICE

As mentioned earlier, glaciers have influenced the distribution of Adélie penguin colonies by transporting nesting stones. In addition, a curious characteristic of Adélie penguin colonies becomes evident where this species nests in the company of its two closest relatives: the chinstrap and gentoo penguin.[49,52] Whereas the latter nest on slopes and ridges, respectively, the Adélie penguin colonies tend to be denser and farther from landing beaches. The reasons for these differences have been attributed to the fact that, on the largest geographic scale (throughout Antarctica), Adélie penguins nest much farther south than the others, and they need windswept spots where snow does not accumulate.[22,49] I contend that another environmental factor has contributed to this Adélie penguin attribute, and it is related to factors more important during periods of maximum glacial advance (LGM, 19,000 y.b.p.). During such periods of ecological stress, a species's attributes are put to the test.

During the glacial maxima, and even today in the southernmost areas where this species nests, almost any ice-free terrain had to be a hill high enough that the ice cap, or runoff glaciers, diverted around it, as occurs today at Mount Biscoe, MacRobertson Land (fig. 3.5). Moreover, during the glacial maxima, the ice cap was so large (and heavy) that it pushed the continent many meters lower than it is today.[7,8] Any gradual coastal terrain was quickly submerged. Very few, if any, locations on the continent were not covered by ice during the glacial maxima, and during the last one perhaps only Cape Adare was exposed. The species had to exist farther north, where pack ice then reached to the very steep, volcanic islands such as Heard, Gough, and Crozet. As the glaciers retreated, the first continental terrain freed for colonization by Adélie penguins was the steep, coastal hills and mountainsides similar to Mount Biscoe. In the northern part of the Adélie penguin's range, where colonies are very small and there

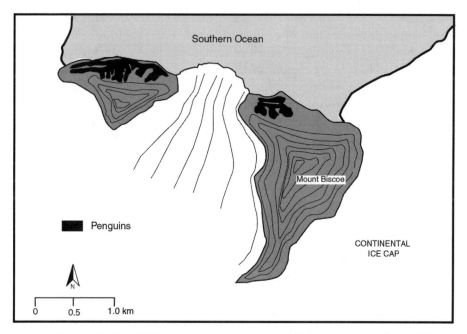

Figure 3.5 The Adélie penguin colony at Mount Biscoe, East Antarctica (66°13′S, 51°22′E), demonstrates the precarious position of many colonies around Antarctica: a projection of land high enough to escape the continental ice sheet but nevertheless bisected by a glacier. These hills force the East Antarctic Ice Sheet to diverge to leave an area ice free.

is no premium on Adélie nesting habitat, its relatives (chinstraps and gentoos) only recently have colonized overlapping locations,[36] thus filling the lowlands beneath Adélie penguin colonies. Near the continental margin (as opposed to the peninsula), where nesting habitat is much more in demand by this species, many Adélie penguin colonies extend down hillsides almost to the water's edge.

SEA ICE

In many places around Antarctica, especially where there are bays or sounds (i.e., stretches of water extending inland from the open sea), fast ice is a common feature. As discussed in chapter 2, fast ice is sea ice that remains in place year-round or almost year-round as it is locked in place by topographic features (capes, islands, grounded icebergs) and protected from ocean swells. Where fast ice is persistent well into the nesting season and wider than 2–3 kilometers from beach to ice edge, Adélie penguins do not tend to establish colonies. Where they do so in such a situation, as at Hukuro Cove, Enderby Land, colonies are very small, many fewer than 1,000 pairs (table 3.1; see also Levick's quote in regard to Duke of York Island, ch. 2). In those situations, penguins usually have to enter the water through a limited number of tide cracks in the fast ice.[50,51] Thus foraging area is limited (by the breath-holding capabilities of the penguin) even if food supply is not.

The classic example of the effect of fast ice on the occurrence of Adélie penguin colonies is provided by McMurdo Sound, which extends southward from the southwestern corner of the Ross Sea (fig. 2.4). The sound is formed between Ross Island and Victoria Land, with the Ross Ice Shelf as its southern shore (an ice shelf is a portion of the continental ice cap that is floating on the sea). Adélie penguins are common in the immediate vicinity; in fact, about 8 percent of the world population nests on Ross and Beaufort islands, with the latter at the northern end of the sound. Except for the penguins nesting at Cape Crozier, all remaining individuals in this complex nest along the northeastern shore of McMurdo Sound (fig. 2.14). Around the remainder of the sound, that is, around its southern and western shores, exists persistent fast ice (fig. 2.4). During most years, the fast ice remains in place year-round, but every four to five years it disappears completely. It is 10–20 kilometers wide at its usual extent but sometimes more and sometimes less.

In the northeastern portion of the sound, the penguins nest everywhere that there is a beach and where no fast ice remains in place beyond about November 15. There are dozens of beaches in the southern and western parts of the sound, yet there are no Adélie penguin colonies. However, along this section of coast are eleven localities where South Polar skuas nest.[4] The point is that suitable nesting terrain is present and was in use a few thousand years ago. The beaches and ice-free terrain of Cape Evans and Winter Quarters Bay were the sites of famous British exploratory expeditions in the early 1900s. These sites were accessible long enough in certain years to allow close access by ship. The fast ice edge in McMurdo Sound usually lies at Cape Royds, site of the world's southernmost Adélie penguin colony and that of another past British expedition's quarters. Just 3 kilometers farther into the sound from Cape Royds is Cape Barne, 3 kilometers of fast ice where Adélie penguins have nested sporadically. In fact, in only one year since 1957 have they nested there: 1987,[44] during a three-year period when the fast ice disappeared very early (by November). The penguins set up territories but did not breed in

1997, the second year of a two-year period of minimal fast ice (ice edge at Cape Barne). With the return of the usual ice conditions in 1998, the Adélie penguins showed no interest in Cape Barne. Enough nesting habitat exists at Cape Barne for many thousands of penguin pairs. More than forty skua pairs nest there.[4] Therefore ice conditions would have to be favorable enough to encourage many more than the twenty to thirty pairs of penguins that have recently shown interest.

Why a walk of 2-3 kilometers over sea ice dissuades Adélie penguins from establishing a colony is unknown. More than half of the Victoria Land coastline is imprisoned by fast ice, and there are no Adélie penguin colonies despite the availability of suitable nesting terrain. The critical time for fast ice to disappear probably is the chick-provisioning period. During that period, Adélie penguins carry huge loads of food, ten times more than they do at any other time of year. At Cape Crozier, for instance, Adélie penguins walk almost a kilometer inland to reach the most distant subcolonies, but where the terrain becomes especially steep they have not spread farther. This factor of energetics may limit the size of Adélie penguin breeding colonies.

Thus it is apparent that the presence of sea ice both encourages and discourages the presence of Adélie penguin colonies, depending on the amount and form of sea ice. Obviously, 100 percent cover (fast ice) during the nesting season is too much. In the Antarctic Peninsula region and on the nearby South Shetland, South Orkney, and South Sandwich islands, where 40 percent of all Adélie penguins colonies exist, sea ice is present in the winter and spring in concentrations approaching or reaching 100 percent cover. However, its presence is short lived. It begins to disappear by late September and October and is usually gone entirely by late November, not to appear again until May or June.[17,45,59] In this area, Adélie penguins have little trouble negotiating ice-choked seas. In fact, a gradual disappearance of sea ice in this region, manifested by a lengthening of the ice-free season (decreased overlap with the Adélie penguin nesting season), appears to have caused Adélie penguin populations and numbers of colonies to decrease.[15,36] This topic is discussed further in chapters 8 and 9.

POLYNYAS

Around the coast of continental Antarctica, polynyas make Adélie penguin colonies possible in many areas where they otherwise could not exist because of the presence of consolidated pack ice well into the nesting season (fig. 3.6). "Polynyas are recurring regions of either partially or totally ice free ocean occurring within the pack ice at times when the air temperature is below the freezing point of seawater" (Van Woert, p. 7753).[48] Polynyas were discussed in chapter 2 as they relate to migration. Polynyas have been studied a great deal in the Southern Ocean but, until recently, only from the physical standpoint. Coastal, latent-heat

Figure 3.6 The Ross Sea and Terra Nova Bay polynyas are formed by persistent winds blowing off the Ross Ice Shelf and continent, respectively. The Ross Passage polynya is formed by upwelling of warm water. Penguin colonies are located along the southwest corner of the Ross Sea polynya, in Terra Nova Bay, and adjacent to the Ross Passage polynya. *Satellite image courtesy of NASA.*

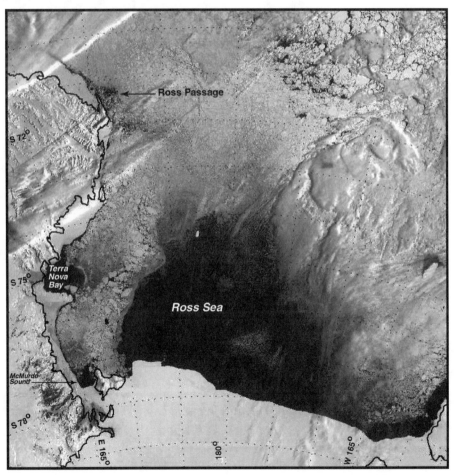

polynyas are formed from the heat expelled from freezing seawater and strong katabatic, offshore winds that push the forming ice away from the coast and are the source of Antarctic Bottom Water. After reaching the bottom, this high-salinity water (the salinity comes from salt rejected in the freezing process at the surface; see ch. 2) flows north into the oceans of the Northern Hemisphere. There it is upwelled and figures importantly in maintaining the global salt budget and ocean currents. Antarctic polynyas therefore are an important factor in global climate. The best-studied polynyas appear to be those in the Ross Sea: Ross Sea Polynya, Ross Passage Polynya, Pennell Polynya, and Terra Nova Bay Polynya (fig. 3.6).[21,43,48] Many books on physical ocean processes in the Antarctic contain discussions of these and other polynyas.[11,20]

The association between seabird breeding colonies and polynyas is a well-known phenomenon in the Arctic.[38] In that region, even for birds of flight that can travel hundreds of kilometers easily, polynyas allow access to the ocean-and food. Stonehouse[41] was the first to recognize the importance of polynyas to seabirds, specifically penguins, in the Antarctic. He hypothesized (I believe correctly) that the colonies at capes Royds and Bird, Ross Island, owed their existence to a polynya in McMurdo Sound. That polynya is an extension of the much bigger one called the Ross Sea Polynya (table 3.1, fig. 3.6), separated from the latter only by Ross Island but generated by the same wind regimes. In other words, the southerly katabatic winds divide when they meet Ross Island, most blowing around the east coast and the remainder blowing around the west coast of the island across McMurdo Sound. Earlier Stonehouse[39] proposed that the Adélie and emperor penguin colonies at Cape Crozier, Ross Island, benefited greatly from what is now called the Ross Sea Polynya. Without these polynyas, the penguins at these colonies would have very difficult access to food. This was confirmed dramatically by studies at Cape Crozier, which showed a marked reduction in breeding productivity in years when the polynya did not develop soon enough in the nesting season.[2] In such years, many penguins cannot travel fast enough between the colony and the open sea to feed and then return to relieve their mates of incubation or to feed their chicks. Therefore eggs are deserted by hungry mates or chicks starve.

Stonehouse's insight went unrecognized for more than thirty years. Little was heard in this regard until 1996, when clusters of Adélie penguin colonies centered around southern Adelaide Island and another around southern Anvers Island, Antarctic Peninsula (Bellingshausen Sea), were analyzed.[46] It was hypothesized that polynyas in Marguerite Bay and over the Palmer Deep, respectively, accounted for the existence of those colony clusters. These polynyas would have to be sensible heat polynyas, which are formed by the upwelling to the surface of warm, deep water. This warmer water retards the formation of sea ice. In fact, Stonehouse[40] earlier proposed this same connection in regard to Adélie penguins on Adelaide Island and a polynya in Marguerite Bay. Subsequently, it has been shown that warmer water does upwell to the surface in these locations.[27,30] Most recently, a link has been proposed between the presence of emperor penguin colonies and latent heat polynyas in East Antarctica.[26]

Farther north than 64°S on the west coast of the Antarctic Peninsula, Adélie penguin colonies are uncommon and small. This probably results from inconsistency in the presence of sea ice after October. In the northern part of the Antarctic Peninsula it has been shown that Adélie penguins have much greater breeding success and larger breeding populations in years when sea ice is much more prevalent than it normally is in that vicinity.[45,47] This is the opposite relationship exhibited by Adélie penguins farther south in their range, where sea ice can impede breeding efforts.[36]

In regard to the seasonal presence of sea ice in the northern portion of the Antarctic Peninsula, it has been proposed that Adélie penguins do not nest on the outer coast of the South Shetland Islands because no polynyas occur there; in contrast, they do nest on the inner, Bransfield Strait coast because open waters are present.[46] It seems more likely that Adélie penguins do not nest on the outer coast because sea ice is absent for almost the entire breeding season, in contrast to the inner coast, where sea ice persists well into the spring each year.[17,59] The eastern coast of the Antarctic Peninsula is a different matter. As noted earlier, except at its tip, no Adélie penguins nest along most of its eastern coast because of heavy ice concentration and a lack of polynyas.

One final word about the basic polynya types is in order. Oceanographers generally concern themselves just with latent heat and sensible heat polynyas. Moreover, by the definition of polynya that these scientists use,[48] polynyas cannot exist past the time of the year when the sea ceases to freeze. In many parts of the Antarctic, Adélie penguins travel to their colonies after this point is reached, yet open water surrounded by compacted sea ice near to colonies is an important part of their migration. In almost all cases, the various latent heat and sensible heat polynyas attractive to oceanographers for other reasons are continued into late spring by these open-water areas attractive to the penguins. Therefore the penguins are defining these open-water areas as polynyas in the old, Russian sense of the word. Naturally, I follow the penguin definition.

By my count, at least 100 Adélie penguin colonies, or 62 percent of the total (161), are known to be associated with a polynya. If one excludes the forty-one colonies located in the Scotia Sea and Drake Passage (South Shetland, South Orkney, South Sandwich islands, and northern tip of Antarctic Peninsula), which are ice free near the start of the nesting season, thus negating a need for a polynya, the percentage comes to 83 percent. Looked at another way, at least 73 percent of the world's breeding pairs of Adélie penguins nest in colonies associated with a polynya; the percentage increases to 75 percent if we exclude colonies of the Scotia Sea and Drake Passage. The latter two percentages (83 and 75) are close, indicating that in the icebound portion of Antarctica most colonies not associated with a polynya are very small. This is evident in table 3.1 and fig. 3.1.

The conclusion is obvious: Coastal polynyas, defined as open-water areas within the pack ice, are critical to the breeding distribution and life history patterns of the Adélie penguin. The amount of sea ice is critical, too. Sea ice persistence and polynya development are sensitive to climate change in both the long and short term. The next three chapters explore the effects of sea ice presence or absence in the short term (annual breeding seasons); chapter 9 looks at the long-term effects on Adélie penguins with respect to changes in sea ice cover.

Chapter 4

The Annual Cycle

The annual cycle of an Adélie penguin includes a premigratory period of feeding and fattening, spring migration to the colony, nesting, fall migration from the colony, continued heavy feeding, and then molt. All this takes about six to seven months; the remaining months—fall and winter—are ones of little activity. The subject of wintering was considered in chapter 2, and nesting is taken up in greater detail in chapters 5 and 6. In this chapter we consider the annual cycle, the timing of its various parts, and molt in some detail. As we shall see, the central part of the annual cycle—nesting—is timed to take advantage of the seasonal presence of sea ice and a seasonal appearance of abundant food. The nesting cycle ends when the sea ice disappears, and molt then occurs after migration to areas where the sea ice is still present. Timing of the cycle is affected by latitude (through a photoperiod effect) and by age and experience.

After four decades of intensive research by dozens of researchers all around Antarctica, we now know much about the breeding behavior and success of Adélie penguins. Much less was known when William Sladen began his research at Hope Bay and Signy Island in the late 1940s and early 1950s, alone and armed only with the earlier findings of Levick, Gain, Bagshawe, and other naturalists from the heroic age. His intuition then was remarkable, judging from what we have learned since:

It soon became clear that the social structure of an Adélie penguin community could not be understood by the study of just a few marked and dissected birds. The larger the sample . . . the more variations in behavior became apparent and the more difficult it was to establish a simple and clear story. At Signy, for example, adults started to return after the winter during the first week in October, although the population did not reach its maximum until one month later. . . . The first eggs were laid on October 29th, yet on November 19th fresh eggs were still being found. Some birds went about their business with an assured manner, others loitered on the periphery of the colonies, or wandered far away from the breeding areas. As I became more familiar with the birds, it appeared that these variations, and other differences in behaviour and breeding efficiency, were due to three important and often interrelated factors. These were, variation among individuals, differences in "intensity" of behaviour, and differences in age and breeding experience. . . .

It was not possible to study Adélies of accurately known age at Hope Bay or Signy . . . , so that the age groups given below are provisional. . . .

Established (experienced) breeders three to five years old and over; mostly four years old and over.

Unestablished (inexperienced) breeders two to four years; mostly three years old.

Non-breeders ("Wanderers") in adult plumage two to three years old; mostly two years old.

Non-breeders ("Wanderers") in immature plumage (Yearlings); from leaving the rookery until they moult into adult plumage when about fifteen months old.

Nestlings (Chicks); up to the time when they leave the rookery, nearly two months after hatching. (p. 25)[45]

Sladen also expressed the hope that other researchers would follow to confirm his guesses using the penguins he was then marking, many as chicks. This did not happen in the Falkland Islands Dependencies where he worked but did occur some years later, especially at Cape Crozier, on the other side of the continent, mostly as a result of his own efforts.

■ Colonial Nesting: Constraints and Advantages

Like the majority of seabird species, the Adélie penguin is a colonial breeder. This means that Adélie penguin breeding territories are closely spaced and that they abut at least one other territory, usually more. Colonial breeding in seabirds, as in other such species, has arisen for two main reasons. The first is limited choices of where to nest because of limited breeding space. The second is a need to congregate for social reasons, such as social facilitation of nesting behavior, increased ease in finding mates, and transfer of information on where to find food.[51,52] Finding food is particularly important for species such as the Adélie penguin that, when away from breeding colonies, dwell at low densities over vast areas (thousands of square kilometers). Moreover, searching for food is much more energetically costly for a penguin than for a flighted seabird.

For most seabirds, nesting habitat is limited by a need for areas free of terrestrial predators. The limit is stricter for Adélie penguins because such areas also must be adjacent to ice-covered seas and must be free of land ice and snow. These sites also must be accessible from the sea by walking. Less than 5 percent of coastal Antarctica is free of glacial ice, and much of that area consists of precipitous cliffs reachable only by flighted birds. The distribution of Adélie penguin breeding colonies is reviewed in chapter 3, including an analysis of factors that affect their occurrence.

Colonial species, such as bees and swallows, learn from others in their colony about where food is most likely to be available. These species, like Adélie penguins, are known as central-place foragers because their feeding forays radiate outward from the colony. They use the direction they see others going to or coming from to narrow down the choice of where to search for food. Otherwise, they would have to return to where they last encountered food. Unfortunately, no direct data are available to confirm or rule out any transfer of information between Adélie penguins: likely not—supposition but probably well founded.

Whether group defense against predation encourages colonial breeding in penguins and other seabirds is debatable. Seabirds nest only where land predators do not exist, usually offshore or on oceanic islands normally free of rats, cats, canids, and other mammals. Such sites are limited. When land predators are introduced to seabird islands, the result is either abandonment by the seabirds or decimation of the colony because many seabirds have no defenses against such intrusions. In some cases seabirds do not seem to recognize that they are in danger. Aerial predators, such as gulls and skuas, are a different matter. Aerial predators take advantage of the concentrated victims (colonial seabirds), but these predators usually are colonial also, which increases their vulnerability to others of their own kind. The aerial predators are colonial for all the same reasons that encourage colonial breeding in other species. There are certain raptors that prey on seabirds, but they are able to kill at will or at least often enough for it to seem that way to us. Such predators occur in very low densities, but where they become prevalent, colonial seabirds usually abandon their breeding attempts.

The major disadvantage of being colonial and therefore a central place forager is that a population can exhaust the food supply close to the colony. This subject was discussed in detail in chapters 2 and 3.

■ Timing of Nesting: General Considerations

Typically, Adélie penguins remain at sea in the pack ice during the late fall, winter, and early spring. Probably at about the time of the spring equinox, they begin their annual cycle with a period of persistent feeding, called hyperphagia, which in turn leads to premigratory fattening. This change in behavior and physiology in response to altered photoperiod is common in birds of high latitude.[32,50] The fat reserves accumulated sustain the penguins as they progress across ice-covered seas to reach nesting colonies. The extensive sea ice limits feeding opportunities.

An exception to the usual quiescence of the nonbreeding season characterizes penguins associated with the most northerly breeding sites. At those sites, daylight occurs for a short time even in midwinter (as opposed to blackness farther south), and sea ice is not extensive, thus allowing the

Illustration 4.1 An Adélie penguin sitting on its pebble nest. *Drawing by Lucia deLeiris © 1988.*

penguins easy access to the land. As noted in chapter 2, to navigate and capture food Adélie penguins need at least a few hours of daylight in the areas where they spend the winter. Most breeding sites for this species are well south of the Antarctic Circle, but a few are not. Winter visits by adult Adélie penguins to colonies have been observed at Gorlay Peninsula, Signy Island (60°S); Water-Boat Point, Danco Coast (65°S); and Arthur Harbor, Anvers Island (64°S).[13,36,45] The only breeding localities for this species even farther north are among the South Sandwich Islands (57–59°S). Daylight during winter is even more extensive, so winter visitations must occur there as well. Farther south, beyond the Antarctic Circle, where it is dark during winter, midwinter visits to colonies by Adélie penguins have not been reported. By about the time of the autumn equinox, Adélie penguins have completed the annual molt and thus their annual cycle.

It is thought that most, possibly all seabirds time their nesting cycles to take advantage of a flush of food that becomes available predictably when adults must feed their young and when those young are learning to forage for themselves.[27,28,32] In most cases, organisms of the middle levels of the food web (i.e., seabird prey) become most abundant and available during summer and fall. These organisms have responded to a cascade of events that begin with increased light levels in spring, followed by increased ocean productivity and abundance of the phytoplankton, then spawning among the grazers, and so on. Adélie penguins, which nest only during the spring and summer, respond to this progression by feeding on the grazers and the grazers' predators (krill and fish). Some penguin species nest during the winter: certain populations of the gentoo penguin, a congener of the Adélie; the emperor penguin; and the Fiordland crested penguin (*Eudyptes pachyrhynchus*). For the gentoo and Fiordland crested penguins, presumably, their phenology (especially the timing of nesting) is also a response to greater food availability, which in their case happens to occur in winter. The emperor penguin's nesting cycle is so long (eight months) that they must lay eggs during the winter so that their young fledge and learn to forage in mid- to late summer when food is plentiful, just as Adélie penguins do.

In the Southern Ocean, annual cycles in marine productivity depend critically on light levels in three ways. First, there is little or no sunlight in winter. Photosynthesis and therefore growth and reproduction of phytoplankton (microscopic plants that drift at and near the ocean surface) cannot happen. Second, in summer, when the sun is above the horizon, the angle of its rays is so low that light penetration of the water is lower than at lower latitudes (where the sun's rays are closer to vertical). In other words, the sun's rays do not penetrate very far, especially in late fall, winter, and early spring. This further shortens the period during which the sun can stimulate photosynthesis. Finally, the sun's rays are blocked by sea ice. In the Southern Ocean, because of the marked pulse of phytoplankton production during summer in response to light, almost every mesopelagic or epipelagic animal species (those occurring at middepths and surface depths) reproduces at that time.[46] The seas are filled with potential prey for penguins. Little wonder that Adélie penguins breed at that time.

That Adélie penguins breed during spring and summer is an adaptation brought about through evolution. Adélie penguins that breed too early probably are not able to find enough food for their chicks; those that breed too late miss the peak in food availability. Only those that time their breeding appropriately are likely to produce offspring with a good chance for survival, thereby ensuring the survival of their genes.

What triggers Adélie penguins to begin breeding—initiation of the so-called breeding urge—is the lengthening of days in spring.[32,50] Presumably, as in other migratory birds, the altered photoperiod first stimulates the penguins to feed voraciously and to accumulate large stores of fat. This hyperphagia begins at about the time of the spring equinox. Body mass probably increases to a level at or near to the maximum for the year, although data on body mass are available only for individuals that have just arrived at the colony after the migration (fig. 4.1). The lengthening days stimulate secretion of certain hormones, which bring on the change in behavior.[50] Further increases in day length and the attainment of breeding condition (fat stores sufficient to sustain breeding) bring on the migration from wintering areas to the colonies. This accumulated fat allows the penguins to migrate, in large part by walking, over seas frozen too solid for ready access to water and food. Moreover, the stored resources give both sexes the time necessary to court and establish or renew their pair bond. The female is able to form the eggs, and the male is able to set up the territory and incubate the egg for the first week or two. Therefore the premigratory hyperphagia and fattening are critical.

This relationship between day length, the appearance of food, and the onset of the annual cycle in the migratory Adélie penguin contrasts with the pattern exhibited by the nonmigratory Galápagos penguin (*Spheniscus mendiculus*), which nests at the equator. There, no seasonal change in

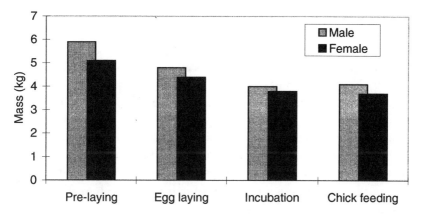

Figure 4.1 The change in body mass of male and female Adélie penguins during the nesting cycle at Cape Crozier, 1969–70. *Data from Ainley and Emison.*[3]

day length is evident. This penguin initiates breeding when there is a precipitous drop in sea surface temperature, regardless of time of year.[14] Such an event signals that a productive marine food web will soon follow and that the probability of finding food for chicks will be high.

In the late summer, when day length begins to shorten dramatically, this change in photoperiod stimulates a cessation of an Adélie penguin's breeding urge and migration away from the breeding colony without reducing the voracious foraging needed to feed chicks. At high latitudes, Adélie penguins must complete their postbreeding migration before the molt. In this way the penguins avoid being trapped where the sea freezes solid to the horizon beginning in late February, not long after the end of the breeding season. With continued foraging, adults accumulate enough body fat to fast through the molt and to form the new feathers. During the molt, which is much more rapid in penguins than in other birds[53] but nevertheless lasts a few weeks, Adélie penguins do not enter the water and therefore cannot capture prey. By the autumn equinox, the postbreeding migration and molt are completed. This pattern has been studied extensively in domestic chickens and other birds.[50] In chickens and other birds, the photoperiod can be managed artificially (by programmed light regimes) to extend the egg-laying period and forestall the onset of molt. In aquariums, where captive Adélie penguins are held, light levels are regulated during the year to mimic the conditions found by these birds in the wild.

■ The Pattern of Colony Occupation

The phases of the Adélie penguin nesting cycle at the colony were first described by William Sladen.[45] The occupation period includes arrival, pairing, egg laying, and most of the incubation period. The colony reaches its maximum size during this period because almost all eventual breeders arrive and pair synchronously. The fact that the breeding cycles of all the penguins in a colony are so closely timed leads to the well-defined phases (fig. 4.2). Among seabirds that breed much less synchronously, such as species in which egg laying could occur in almost any month of the year (tropical and equatorial species such as African and Galápagos penguins), discrete phases of the breeding cycle within the population are difficult to discern.

The first Adélie penguins to arrive at a colony in the spring are males and, on average for established pairs (those that nested together the previous season), the male usually arrives earlier than his mate. During the 1968–69 nesting season at Cape Bird, Ross Island, the males arrived four days earlier on average than their mates of the previous season; only 4 of

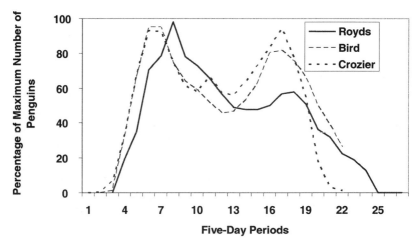

Figure 4.2 The annual cycle of occupation by Adélie penguins at capes Royds, Bird, and Crozier, Ross Island. Data are expressed for five-day periods as the percentage of the number present when the annual highest count occurred (in early November). For each site, data from multiple years are averaged: three years at Royds during the 1950s and 1960s,[48,58] two years at Bird during the late 1960s,[47] and three years at Crozier during the late 1960s-mid 1970s.[5,9]

133 females returned before their respective mates.[47] Once eggs are laid, the females leave to feed at sea for eight to fourteen days (depending on the extent of sea ice). As more females complete egg laying, the number of birds present begins to subside. Few birds arrive at this time to initiate nesting. Once all eggs are laid, mostly just the males that are incubating eggs remain. Some eventually lose their eggs and leave, too. At this time the colony population dips to a level that is about half the occupation period peak (exceptions are discussed later in this chapter). Mates change places on the eggs, but the nonincubating bird leaves quickly for sea in search of food. This is a good time to census an Adélie penguin colony to estimate the number of nesting pairs because just one member of each pair and only a few nonbreeders are present.[49]

Soon after the low point in colony population is reached, young nonbreeders begin to arrive to prospect for territories and practice social interactions (see ch. 5). Also arriving then but actually returning to reoccupy their territories are failed breeders. These birds have been gone for a couple of weeks, during which they have restored their body fat. These birds will remain for a few weeks. Therefore the reoccupation period is named for these returning failed breeders. The colony population grows toward a second peak, usually not as great as that reached during the occupation period (exceptions are discussed later in this chapter). The population during the reoccupation peak is composed of breeders, unsuccessful breeders (who are reoccupying their nests), nonbreeders, and young birds prospecting for nests.[45] Thereafter, numbers of adults begin to decline as more and more breeders leave their chicks, who in turn band together to form crèches. Eventually, the nonbreeders leave as well. Remaining are the large chicks and the parents who visit briefly to feed them. Once the chicks fledge, a few molting adults can be found somewhere in the colony area. Otherwise, what once seemed ordered chaos is now silent.

This pattern in seasonal population dynamics differs depending on location. On Ross Island, the pattern was quantified at three different colonies during the late 1960s and early 1970s (fig. 4.2). The data presented here have been averaged for several seasons at each colony. At capes Crozier, Bird, and Royds, the synchrony of penguins during the occupation period is similar, as indicated by the shape of the population curve. At capes Crozier and Bird the timing of the occupation period is identical but precedes that of Cape Royds by five to ten days. The delay at Royds probably results from the much more extensive fast ice that covers

McMurdo Sound in early spring. Penguins usually have to walk up to 35 kilometers over this ice and also negotiate heavy pack ice seaward of it.

In contrast to the occupation period, the reoccupation period differs greatly between the three colonies. At Cape Crozier, the reoccupation peak is almost as great and as sharp as that of the occupation period. The reoccupation peak at Bird is lower, and that of Royds is much lower. The reason for these differences is not clear. The Crozier colony is an order of magnitude larger in population than that at Bird, and Bird is an order of magnitude larger than Royds. Also, Crozier, then Bird, is closest to the Ross Sea Polynya, so extensive open water is available early in the occupation period. Therefore travel to Crozier is the easiest. One possibility is that the larger size and easier access of Crozier attract potential recruits from other colonies who may be exploring for territories during the reoccupation period. These birds get caught up in the dense traffic going to and from Crozier and find themselves ashore there, hence the larger peak of prospectors. Another possibility is that nesting success decreases from Cape Crozier to Bird to Royds, so more birds reoccupy failed nests at Crozier and Bird than at Royds. By attending failed nests rather than just visiting briefly to feed chicks, it is more likely that these birds are counted. Although this latter possibility would explain the pattern, there is no evidence that it is true. The first explanation seems more feasible, especially during a period when sea ice was more extensive (than now) and these colonies were declining (Royds faster than Bird, which declined slightly faster than Crozier). Under the more difficult ice conditions, the more accessible Crozier colony probably would be more favored by immigrants.

Other differences between colonies in the timing and shape of their seasonal population curves are apparent when viewed Antarctic-wide (fig. 4.3). In this case, numbers for the three Ross Island colonies have been averaged together. There is very little difference in the population dynamics of colonies at Point Géologie (66°40′S), the Windmill Islands (66°20′S), and Ross Island (77°20′S). In contrast, the colony occupation cycle is advanced at Signy Island (60°40′S) by ten to fourteen days and is retarded by a similar amount at Béchervaise Island (67°30′S). Several factors might explain the differences. First, Signy is much farther north, so the photoperiod is advanced. Also, Signy Island is free of consolidated pack ice several weeks earlier than the other sites mentioned (October as compared to December, January, or later at the others). In short, access is easier earlier. Third, Béchervaise Island is entrapped in extensive fast ice for all or most of the breeding season. Penguins usually have to walk

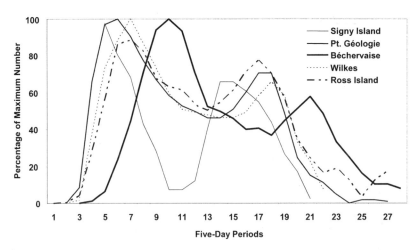

Figure 4.3 The annual cycle of occupation by Adélie penguins at five localities around Antarctica. Data are expressed for five-day periods as the percentage of the number present when the annual highest count occurred. For Ross Island, data from eight years at three colonies are averaged (from fig. 4.1); data for other localities are from single years.[26,38,43,45]

10–25 kilometers between the colony and open pack. Even more so than Cape Royds, nesting may be retarded at Béchervaise because of the greater effort needed to travel to the colony from the wintering area. This tends to spread out the arrival of penguins. Finally, the population dip between the occupation and reoccupation periods is far more extreme at Signy Island than at any other site where colony counts are available. This seems to argue for an even greater degree of breeding synchrony than elsewhere, perhaps facilitated by more synchronous arrival. The latter in turn might result from ease of access stemming from the loose pack ice present from very early in the spring. In other words, the presence of pack ice tests the vigor and desire of individual penguins, with some coping more easily than others. Pack ice is gone by the start of October in the vicinity of Signy Island but not until about January 1 at Ross Island and maybe not at all at Béchervaise. On the other hand, the early, synchronous breeding at Signy may represent a rush by these pack-ice-obligate penguins to undertake breeding before the sea ice disappears. Sea ice is absent within hundreds of kilometers of Signy Island by the time Adélie penguin chicks fledge.

■ Timing of Nesting: Geographic Variation

Adélie penguins find suitable nesting habitat in coastal areas spanning about 20° of latitude (58–78°). That is equivalent to the geographic span from central Baja California, Mexico, to Anchorage, Alaska, or from northern Africa to Norway. Few other coastal avian species breed over that extensive a geographic span, particularly among those (such as Adélies) that exhibit no geographic variation in morphology (and consequent divisions into subspecies). The exceptions are seabird species that breed in the tropics (a broad climatic band encircling the middle of the earth and bounded by the tropics of Cancer and Capricorn) and experience little seasonal variation in day length or climate. Exceptions also are terrestrial cosmopolitan species such as the English sparrow (*Passer domesticus*) and European starling (*Sturnus vulgaris*), which have colonized almost everywhere that humans have colonized. Given such a wide extratropical north-south geographic span to the nesting range, with a corresponding range in climate and light regimes, does breeding occur among all populations of Adélie penguins at the same time? Not surprisingly, the answer is no.

Figure 4.4 The date in spring when the first Adélie penguins arrive at various colonies, by latitude (degrees south). *Data from various sources.*[5,26,31,38,40,45,48,54]

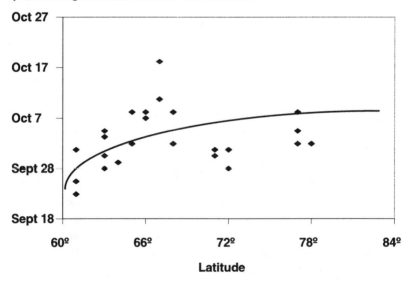

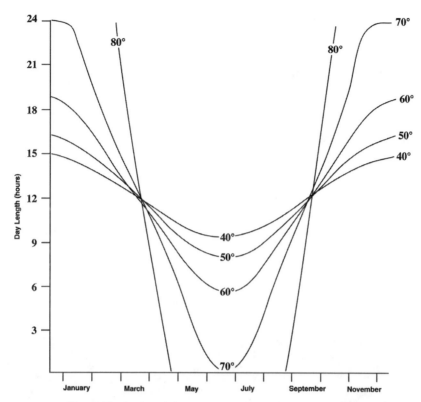

Figure 4.5 Changes in day length by date at various latitudes in the Southern Hemisphere. *Redrawn from Smith.*[46]

As mentioned earlier, premigratory hyperphagia and then migration initiate the annual breeding cycle for Adélie penguins. At the lowest latitudes (and where data happen to be available, as at Signy Island), nesting site arrival on average occurs almost twenty days earlier than at colonies of greatest latitudes (Ross Island; fig. 4.4). Actually, the delay in arrival increases steadily with latitude from 60°S (Signy) to about the Antarctic Circle (66°30′S), with a difference of about twelve days over the 6.5° latitude change. Farther south, across the next 12° of latitude in which this species nests, there is much less difference in arrival date (about eight days). The very rapid seasonal change in day length south of the Antarctic Circle probably has something to do with this pattern (cf. fig. 4.5).

Variation in the relationship between timing of migration and day length probably also has to do with the extent of sea ice in spring. Where sea ice is habitually extensive and compacted—Ross Island and Béchervaise Island—arrival seems to be even more retarded. Where Adélie penguins winter close to the nesting colonies and spring polynyas are present—Signy Island, Hope Bay, and capes Adare and Hallett—arrival is advanced.

■ Timing of Nesting: Effect of Age

FIRST RETURN TO NESTING COLONIES

A few Adélie penguin yearlings, distinguishable by their white chins and throats[45] (fig. 4.6), return to colonies each year in January and February. A few of these one-year-olds rest near or wander among the subcolonies, but most arrive to molt after the nesting season is finished. Year-

Figure 4.6 A yearling Adélie penguin, most of which do not visit colonies until a year later. This bird is beginning its molt.

lings account for much less than 1 percent of all birds visiting a colony each year. By contrast, large numbers of yearling chinstrap and gentoo penguins visit their respective colonies each year, especially to molt. The reason for the difference probably is related to the migratory nature of the Adélie penguin and its association with sea ice. Sea ice, the favored marine habitat of the Adélie penguin, usually has retreated a significant distance from Adélie penguin colonies by February, the period of molt. The yearlings retreat with the ice.

In contrast to the situation ashore, many one-year-old Adélie penguins are found in pelagic waters in the vicinity of colonies during December and January, as long as sea ice is present. Although they do not visit colonies, these youngest birds have moved toward breeding areas and have left vacant large portions of seemingly suitable habitat (Fig. 2.11). By January, they spread out to reoccupy all of the available pack ice and presumably concentrate where food is abundant. Perhaps the mass movement of older birds toward colonies during the reoccupation period draws these youngsters along or, stimulated by a slight increase in hormone levels, they exhibit an incipient migratory movement toward colonies.[1,7]

Adélie penguins do not begin to visit colonies in large numbers until two years of age. By the time they are three years old, 72.0 percent of males and 80.4 percent of females have visited at least once (table 4.1, fig. 4.7). The average age of first return as determined at Cape Crozier is 3.1

Table 4.1 Age of Adélie penguins at first visit to Cape Crozier, 1963–1969.

Age (yr)	Male Percentage	n	Female Percentage	n
2	28.6	157	30.7	158
3	43.4	238	49.7	256
4	18.9	104	14.6	75
5	7.5	41	3.3	17
6	1.6	9	1.0	5
7	0.0	0	0.6	3
8	0.0	0	0.2	1
Total	100.0	549	100.1	515

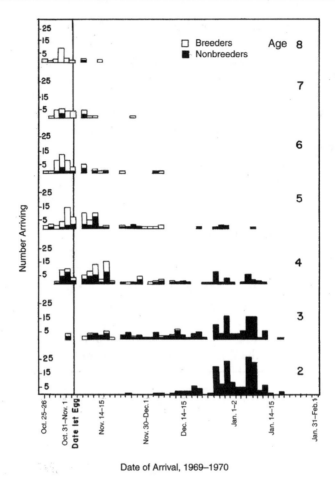

Figure 4.7 The number of Adélie penguins by age group arriving each day at Cape Crozier, 1969. *From Ainley et al.*[6]

years ($SE = 0.01$) in males and 2.9 years ($SE = 0.08$) in females ($P < .05$, t test). By the age of four years, 90.9 percent of males and 95.0 percent of females have visited a colony. Almost all Adélie penguins have visited a colony by age seven. The slightly earlier return of females probably is related to their earlier reproductive maturation.[1] Females recorded in table 4.1 as making first visits when six to eight years old, probably were missed as younger birds. These older females bred in the year first seen, but it is unlikely that they bred without having visited in a previous season. It is also possible that they first visited a nearby colony.

ARRIVAL DATE IN THE COLONY

Older Adélie penguins arrive at colonies earlier in the spring than younger birds, and individual birds usually arrive earlier in a given spring than they did the previous year (table 4.2). These patterns continue only up to a certain age, depending on sex, and are related to whether a bird arrived for the first time or whether it had visited during a previous year, as well as its breeding status. The date the first bird of any given age class is seen varies from year to year and does not follow a precise age sequence among individuals older than five years. This variation is caused by individual variation among the birds composing each age group.[6]

The advances in arrival date with age for males and females are most dramatic through ages seven and six years, respectively; the differences then decrease. Among males seven to eleven years old, arrival dates differ little for birds that had visited in a previous year, but individuals of the oldest age classes arrive even earlier. Among females, there is little difference in arrival dates after six years of age. Breeders in a given season arrive earlier on the average than nonbreeders, a pattern that holds within respective age classes (table 4.3).

The trend toward earlier arrival of older birds is confounded by the tendencies of individuals, some of which return consistently early, some consistently late, and a few at widely varying dates. However, the overall pattern is earlier return with greater age and experience, but with each added year of age the change becomes less (fig. 4.8). Birds two to three years old advance the median date (i.e., the date at which 50 percent have returned) by twenty-one to twenty-five days; from three to four years, sixteen to twenty days; from four to five years, eleven to fifteen days; from five to six years, six to ten days; and from six to seven years, zero days. More than four-fifths of Adélie penguins that are three years of age arrive earlier than they did the previous year, but among seven-year-olds, half arrive earlier and half arrive later.

Ice conditions in spring affect arrival dates of Adélie penguins, a subject discussed in depth earlier. However, the effect of sea ice extent also varies with individual birds' age and experience but not sex. At Cape Crozier, pack-ice cover was extensive in the spring of 1967 and 1968 but minimal in 1969.[5] On average, the date on which the population reached its occupation period peak was a week earlier in 1969 than in the previous years. Arrival dates of two-, three-, and four-year-olds were affected little, probably because the ice was gone or greatly dissipated

Table 4.2. Average date of arrival by age among Adélie penguins at Cape Crozier, 1967–1975.

	Males				Females			
Age (yr)	First Visit	n	Returnee	n	First Visit	n	Returnee	n
2	Dec 31 ± 1.0	104			Dec 30 ± 1.0	115		
3	Dec 18 ± 1.2	183	Dec 4 ± 2.5	61	Dec 14 ± 1.4	173	Dec 1 ± 2.4	55
4	Dec 9 ± 2.1	95	Nov 22 ± 1.3	194	Dec 14 ± 2.6	65	Nov 24 ± 1.5	161
5	Dec 6 ± 3.4	40	Nov 16 ± 0.8	409	Dec 15 ± 6.7	16	Nov 18 ± 0.8	317
6	Nov 26 ± 8.0	9	Nov 11 ± 0.6	518	Nov 22 ± 9.2	5	Nov 15 ± 0.6	408
7			Nov 9 ± 0.6	391			Nov 14 ± 0.6	331
8			Nov 6 ± 0.8	198			Nov 12 ± 0.6	164
9			Nov 7 ± 1.1	78			Nov 13 ± 0.8	46
10			Nov 8 ± 1.6	64			Nov 14 ± 2.0	36
11			Nov 6 ± 1.6	59			Nov 14 ± 2.7	33
12			Nov 3 ± 1.2	38			Nov 11 ± 1.3	24
13			Nov 2 ± 2.1	15			Nov 14 ± 0.7	8
14			Nov 4 ± 6.0	3				
All	Dec 18 ± 0.9	431	Nov 12 ± 0.3	2,028	Dec 17 ± 1.0	374	Nov 17 ± 0.3	1,583
Regression*	−0.9786		−0.9523		−0.8406		−0.9357	

*Regressions of age versus arrival date using weighted averages (all cases, $P < .05$).

Table 4.3. Median dates of arrival among known-age breeders and nonbreeders at Cape Crozier in 1967.*

Age (yr)	Breeders	Nonbreeders
1		Jan 29
2		Dec 30
3	Nov 22	Dec 14
4	Nov 19	Dec 28
5	Nov 17	Dec 23
6	Nov 16	Dec 2

*Regression of age versus arrival date: breeders, $r = -0.9759$; nonbreeders, $r = -0.8847$ ($P < .05$). A comparison of dates for breeders and nonbreeders 3 to 6 years of age indicates that breeders arrive earlier ($\chi^2 = 42.48$, $P < .05$).

by the time they migrated to the rookery in late December (cf. tables 4.2 and 4.4). This pattern became apparent after the arrival date anomalies were averaged for each age class during the years 1967–1969. The standard deviation of the mean anomaly for five- and six-year-olds was much larger than for the two-, three-, and four-year-olds (table 4.4). These more varied anomalies indicate that older birds solved the problem with varying degrees of success. For one thing, arrival dates varied much more for first-time visitors than for those who had visited in at least one previous year. In the years of extensive ice, migration was delayed in all older age groups, especially among inexperienced birds; in the years of sparse ice, migration was advanced (for further details, see Ainley et al.,[6] pp. 29–31).

TIME SPENT IN THE COLONY

An Adélie penguin's age, sex, and breeding status have bearing not just on arrival date but also on departure date. The date of last observation becomes later as follows for nonbreeding birds from two-year-olds to eight-year-olds (and older): January 5–11 for males ($r = 0.9092$, $P < .05$) and January 5–9 for females ($r = 0.7960$, $P > .05$). For breeders the trends for three- to eight-year-olds (and older) are as follows: January 8–10 for males ($r = 0.9449$, $P < .05$), and January 9–11 for females ($r = 0.7042$, $P > .05$). Thus the pattern is statistically significant for males but not fe-

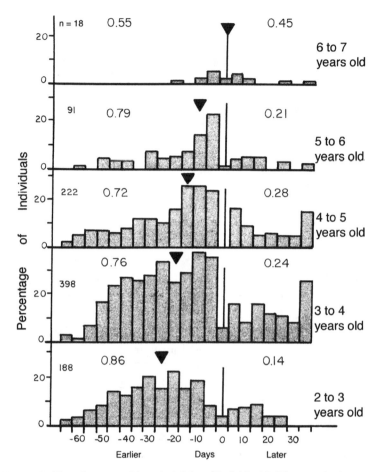

Figure 4.8 The advancement in arrival dates of individual Adélie penguins in successive years, 1963–1969. Each histogram shows the percentage of individuals whose date of return was earlier or later than in the previous year. The percentage of total individuals that advanced or retarded their dates is also shown; arrows indicate the median number of days advanced. *From Ainley et al.*[6]

males. Many of the oldest nonbreeding birds ceased to visit just before or just as the youngest birds arrive (cf. table 4.2). Most of these older birds have been in the colony for two extended periods, the occupation and reoccupation periods, during which they fasted.[2] Among nonbreeders, females depart slightly earlier on average than do males ($t = 4.015$, $P < .05$); no such difference is apparent among breeders.

Table 4.4. Effect of sea-ice extent on arrival date by age, experience, and sex of Adélie penguins at Cape Crozier, 1967–1970.

		1967–68				1968–69				1969–70			
		Male		Female		Male		Female		Male		Female	
Age (yr)	SD	F	R	F	R	F	R	F	R	F	R	F	R
2	1.94	−2		−2		−1		+3		+1		0	
3	2.84	−4	−2	−2	+6	+3	0	+3	0	+4	+1	+1	−2
4	2.86	+2	+3	+1	+3	−5	0	−5	−1	0	0	+4	0
5	12.6	+12	+3	+22	+2	+24	+2	+37	+1	+20	−2	+7	−3
6	10.2	+6	+17	+10	+11	+29	+4		+4	−4	−2		−3
7						+12		+10		+4			−3
8											−1		−10

Data from LeResche.[29]

The number of days before or after the mean arrival date for a given year is compared with the average for all years. Negative numbers indicate earlier and positive ones later than the average.

SD = standard deviation of the mean anomaly for all three seasons; F = first-time visitors; R = returning birds.

The average date when breeding birds are last seen is more a measure of when they discontinue guarding their chicks (which then form crèches; see ch. 5). From then to early February, although parents continue to feed chicks, visits to nest sites are short and easy to miss by researchers (for more details, see Ainley et al.,[6] pp. 31–33).

The time span over which an Adélie penguin visits the colony each nesting season is affected by age, breeding status, and sex (table 4.5). First, older nonbreeding penguins stay in the colony over a longer time period than younger ones. Second, breeders stay longer than nonbreeders, but age does not affect the amount of time that breeders are present. Finally, particularly among nonbreeders, males stay much longer in the colony than females. In part, the first and second patterns help to explain why the peak in colony population during the reoccupation period generally is lower than the peak of the occupation period (fig. 4.2). The varied dates and short-term visits characteristic of younger birds reduce the apparent number of penguins visiting during the reoccupation period. Actually, more penguins visit during the reoccupation period than general counts indicate.

Table 4.5. Average number of days between the dates when Adélie penguins, by age and breeding status, were first and last seen during the nesting season at Cape Crozier, 1967–1974.

	Nonbreeders				Breeders			
	Male		*Females*		*Males*		*Females*	
Age (yr)	Days (mean ± SE)	n	Days (mean ± SE)	n	Days (mean ± SE)	n	Days (mean ± SE)	n
2	4.3 ± 0.6	104	6.0 ± 0.8	115				
3	19.1 ± 1.2	244	17.9 ± 1.3	189			46.7 ± 3.1	38
4	33.4 ± 1.6	256	21.9 ± 2.1	117	48.9 ± 4.0	33	50.8 ± 1.9	109
5	38.6 ± 1.6	270	23.9 ± 2.2	105	52.3 ± 1.7	179	50.2 ± 1.2	227
6	44.9 ± 2.2	128	29.2 ± 3.1	61	60.2 ± 1.1	227	55.2 ± 1.0	221
7	36.1 ± 3.7	46	22.4 ± 4.9	18	63.0 ± 1.1	183	56.2 ± 1.0	179
8	34.9 ± 10.2	10	30.0 ± 10.9	5	62.6 ± 2.2	50	56.3 ± 2.2	4
9	50.0 ± 9.3	9	49.0 ± 7.0	4	60.2 ± 4.9	28	60.6 ± 1.4	18
10					59.5 ± 3.6	34	55.1 ± 2.0	17
11					57.5 ± 4.2	24	45.9 ± 7.5	9
12					72.4 ± 1.9	10	57.8 ± 2.1	10
13					55.8 ± 8.9	4		
Total	30.0 ± 0.8	1,067	18.9 ± 0.8	614	58.4 ± 0.7	772	53.3 ± 0.6	868
r	0.8257*		0.8779*		0.5045		0.4026	

*Regression analysis, $P < .05$.

NUMBER OF VISITS TO THE COLONY

The number of visits a bird makes is also related to its age and breeding status. Most nonbreeders younger than five years of age make only one visit, in December. Older nonbreeders visit twice, arriving in November for about three weeks, leaving for three weeks, and then returning in late December for another three-week stay.[2] Thus the younger birds are present only during the reoccupation period. The older birds are present then in addition to their visit during the occupation period. The pattern is completely different for breeding birds, who make thirty-five to forty visits over the course of a season. For breeders and nonbreeders of both sexes, the amount of time spent in the colony during the first or only visit is inversely correlated to arrival date[2] and to body mass.[1] The average number of days present for breeding and nonbreeding males, respectively, during their first (or only) visit is 26.2 ± 0.3 and 9.8 ± 0.4 days ($t = 37.15$, $P < .05$); among breeding and nonbreeding females, respectively, the values are 10.1 ± 0.2 and 4.6 ± 0.4 days ($t = 13.87$, $P < .05$). Simply stated, if a bird has more stored fat, it arrives earlier and remains longer (see also Ainley et al.,[6] p. 36).

■ The Molt

The final stage in the annual cycle of the Adélie penguin is the molt. All birds molt at least once per year to maintain their plumage. This is especially important for penguins because the plumage is the first defense

Illustration 4.2 Two adult Adélie penguins that have mostly finished their molt. *Drawing by Lucia deLeiris © 1988.*

against the cold water. Moreover, the plumage enables penguins to move effortlessly and quickly through the water (ch. 2). Polar penguins have the densest plumage of any bird, with feather densities reaching forty-six per square centimeter.[53] Another adaptation of molt seen in penguins as a defense against the cold is the process of feather replacement. The new feather penetrates the basal shaft of the old feather, and as the new feather grows, it pushes the old one out, so there is never a gap in the plumage. In most other birds, a feather drops out and a new one appears later.[53] In penguins batches of feathers appear to peel off in sheets.

Considering the molt to be the end of an Adélie penguin's annual cycle assumes that the penguins have migrated from the nesting colony to the pack ice before they molt, which most of them do. For those that molt at breeding colonies, they still must make their postbreeding migration after molting. This migration must be made before the sea freezes near to the colony. If these postmolting birds do not migrate before the sea freezes, they must continue their fast. Because molting takes a heavy toll on a penguin's fat reserves, being trapped by a frozen sea could be fatal.

TIMING OF THE MOLT

The date on which the first penguins at a particular site begin to molt may be affected by day length, as it is in many birds.[50] Day length begins to shorten sooner at the higher latitudes (fig. 4.5), and apparently at higher latitudes the first molting birds are observed at an earlier date as well. For instance, the first molting adults are seen on about February 1 at Cape Royds (77°S);[48,58] February 7 at the Windmill Islands, Wilkes Land (66°S);[37] February 12 at Port aux Phoques, Adélie Land (66°S);[19] and February 10 at Béchervaise Island, MacRobertson Land (67°S).[26] Thus high-latitude Adélie penguins have not only a shorter nesting season but perhaps a shorter annual cycle as well, in part because of an earlier molt.

Actually, the first birds to molt are yearlings, followed by adults (fig. 4.9). However, very few yearling Adélie penguins have ever been observed molting, usually just a few in any year at a particular site. Among the adults, first the failed breeders begin to molt, followed by successful breeders.[37,45] Seemingly, this order is determined by the degree to which a bird was involved with breeding during the month before the molt and when it could break off from its activities to build up the fat reserves

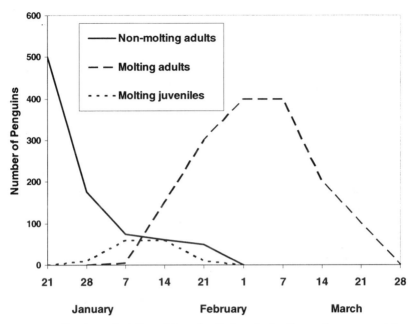

Figure 4.9 Numbers of adults and juveniles that molt after the nesting season at the Windmill Islands, Wilkes Land, Antarctica. *Data from Penney.*[37]

needed during the molt fast. Among some breeding penguins banded at Cape Royds in 1959–60, which eventually molted there, molt was initiated on average nine days after their chicks fledged.[48] This appears to be a much shorter interval than observed at the Windmill Islands in 1959–60, where the peak of fledging occurred during the first few days of February and the peak in numbers of molting breeders occurred during the first few days of March.[37] This indicates a lag of about a month between fledging chicks and beginning the molt. It is possible that food was more available near Cape Royds than it was near the Windmill Islands. Therefore the Royds birds could replenish body condition more quickly.

Unsuccessful breeders at Cape Royds started their molt on average forty-six days after losing their eggs or chicks, although neither the date of loss nor the date that molt was initiated has been specified among these observations.[48] Unsuccessful and successful breeders at the Windmill Islands initiated molt at about the same date on average, March 1–3.

LOCATION OF THE MOLT

Most Adélie penguins molt on the pack ice, not on land.[45] The pack ice has receded far from many Adélie penguin nesting colonies by the time chicks fledge. Migration before the molt helps to maintain contact with the sea ice habitat and precludes the possibility that the sea will freeze solid before or soon after the molt is finished. Captain Worsely reported that the Shackleton party encountered thousands of molting Adélie penguins on the Weddell Sea pack ice between mid-February and late March 1916.[45] Many thousands of molting Adélies have also been encountered on the pack ice of the eastern Ross Sea and adjacent Amundsen Sea during the same time period.[4] On the ice floes, the molting penguins remain in the lee of hummocks; in fact, no molting penguins are seen on flat ice floes. Perhaps the hummocks provide some shelter from winds, or maybe the penguins enjoy the view from a higher perch.

The few penguins that molt on land also seek out sheltered spots. The number of penguins molting on land at a given site, if at a colony, usually is far fewer than the breeding population of that site. Also, the number of birds molting at a colony fluctuates markedly from year to year, presumably as a function of ice conditions in the immediate vicinity.[45] For example, very few molted at the large Cape Adare colony (170,000 pair) in 1899,[23] but a few thousand did so in 1903[55] and 1911.[30] On the other hand, about 800 adults molted at the small Cape Royds colony (1,300 pair) in 1956,[12] compared with about 550 in 1960[48] and 1,370 in 1965 (i.e., about half the number that bred at the colony).[58] Recall that in 1965 the seas around Cape Royds were unusually free of sea ice very early in the summer. This probably encouraged more penguins than usual to venture south. At the Windmill Islands, the number of birds molting was about a fifth of the colony population (1,350 pair) in 1959–60.[37] Finally, only a few hundred, if that, molt at Cape Crozier in any given year (120,000 pair),[9] and in some years none molt at Gorlay Point, Signy Island (10,000 pair).[45]

The Adélie penguins that molt on land don't necessarily seek colony sites or even their own colony sites. For example, in McMurdo Sound several dozen Adélie penguins molt at Cape Evans (12 kilometers south of Cape Royds) or even Hut Point (25 kilometers south) in years when these noncolony sites are open to the sea by late January and February.[48] Taylor[48] marked 208 adults (8 percent of the breeding population) nesting at Cape Royds in 1959–60. Of these banded individuals, only thirteen returned to molt (6 percent of banded birds, 2 percent of molting birds) by late February when the number of molting birds had passed its peak (550) and observations ended. Ten of the thirteen were successful breeders; the other three were unsuccessful breeders. At the Windmill Islands, only 3 percent of banded adults returned to molt at the end of the nesting season.[37] Therefore the majority of molting birds at colonies must be individuals from other colonies or young, nonbreeding birds. Of a sample of molting birds banded in 1958–59, the small proportion that returned to breed at the Windmill Islands the next year were far less successful at breeding than the average adult.[37] The low success rate suggests that the banded molting birds were young and that they bred for the first time in the next year (see ch. 5).

Interestingly, the majority of adults that both bred and molted at the Windmill Islands colony were males (66 percent).[37] Seemingly, males' greater degree of faithfulness to territory (and colony), as exhibited during the nesting season (ch. 5), is expressed also in where they molt.

THE MOLT PROCESS

The molt period begins when a penguin comes ashore and ends when it leaves for sea, having replaced its feathers.[42] Once they have replaced their feathers, the hungry adults leave almost immediately. On average, juveniles complete the molt in 18.6 days (range 15–21) and adults in 19.8 days (range 15–25) on the basis of a large sample of birds observed at the Windmill Islands, Wilkes Land.[37] In short, there is not much difference between juveniles and adults in this regard. In a smaller sample of adults observed at Cape Royds, the duration of molt ranged nine to seventeen days.[48] Compared with other avian species, this is a very short molt. For example, songbirds that live in the Arctic and are considered to molt quickly take forty-eight days to do so.[32] Some tropical seabirds molt through most of the year, changing just a few feathers at a time.

During the molt, the penguin renews all its feathers, but first it feeds voraciously (hyperphagia) to build up its fat reserves. While molting, Adélie penguins do not feed. In fact, just before molting, the penguins seem to be fatter than at any other time of year (fig. 4.10), although no one has ever weighed Adélie penguins just before their southward migration in the spring. This would entail weighing individuals found in the outer reaches of the pack-ice belt. For the spring, only measurements of postmigration and prebreeding body mass are available (fig. 4.1). It is

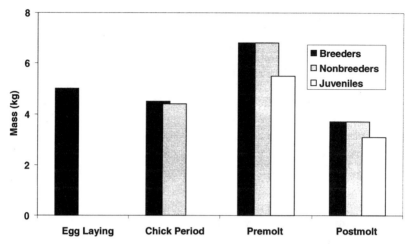

Figure 4.10 The average mass of Adélie penguins (sexes combined) during different phases of the annual cycle; Windmill Islands, Wilkes Land, 1959–60. *Data from Penney.*[37]

likely that the premigration body mass attained during spring equals that of the premolt given that the spring migration takes days or weeks and that body mass loss is very rapid during other fasts (e.g., prelaying and laying).[3,45] In any case, as a result of the molt process and fast, Adélie penguins pass from perhaps the heaviest body mass of the year (premolt) to the lightest (postmolt; fig. 4.10). They lose about 40 percent of their premolt mass,[19,37] so it is not surprising that they seek food immediately after the molt is completed.

The molt also has been described in detail elsewhere.[37] When penguins come ashore (either on land or on ice) to molt, they are very lethargic and irritable. They seek spots protected from the wind and, if on land, near snow banks. They eat a lot of snow to maintain their fluid balance, but otherwise they remain immobile except for preening. They do not enter the ocean. Their plumage initially looks dull and dry, bleached from several months of exposure to continual sunlight. After one or two days, their flippers begin to swell as the size and density of the capillaries in their skin increase. By about the fifth day, the blood vessels in their skin have grown to the extent that they bleed easily if their skin sustains a blow. Their irritability ensures that others do not come near. The enhanced blood supply to the skin brings the nutrients that sustain feather growth. At this time, body feathers begin to stand out and then drop out, especially from the chest and rump. Feathers then begin to drop rapidly all over the penguin's body. About fifteen days later (twenty days after the penguin has come ashore), the only old feathers remaining are tufts on the neck, crown, and legs. Some birds depart at this time; others remain for another couple of days, during which the last old feathers disappear. The new, dark feathers have a bluish sheen. The tail is stubby. Most birds depart with their tail feathers no more than 1–2 centimeters long (eventually to reach about 10 centimeters). Each penguin leaves behind a large pile of feathers.

■ Further Thoughts About the Adélie Penguin's Annual Cycle

Adélie penguins are colonial breeders for want of ice-free terrain on which to set up territories, lay eggs, and raise chicks. As a factor in the colonial nature of breeding, the premium for breeding habitat is demonstrated by this seabird species far better than most. Any snow-free, low-lying terrain adjacent to the sea anywhere around the coast of Antarctica is used for breeding by this species. The other major constraint on nesting, discussed mainly in this chapter, is time.

THE TEMPORAL WINDOW FOR BREEDING

At high latitudes, the Adélie penguin has but a short temporal window in which to complete breeding. It expands that window by being able to fast for long periods. In that way it overcomes the severe ice conditions of spring, which otherwise prevent access to the sea and extend the amount of time needed to travel from feeding areas to and from the colonies. Penguin species closely related to the Adélie (gentoo and chinstrap) do not or cannot fast for long periods and therefore must wait for the sea ice to dissipate before they begin their annual breeding effort. Moreover, they are not as adept as the Adélie penguin in exploiting the sea-ice habitat. A huge portion of the Southern Ocean—that of the sea-ice zone—therefore is unavailable to these two species, which exist around its periphery.

The Adélie penguin also expands its window by undertaking the annual molt in the pack ice after it has migrated away from nesting colonies. The two related species molt at the breeding colony. In part, migrating to molt in the pack ice also enables the Adélie penguin to maintain contact with sea ice, which by the time that chicks fledge (February) has retreated to just a few locations. Ultimately, then, the annual cycle of sea ice determines major aspects of this species's natural history patterns. Proximally, all is mediated by changes in day length, or so it seems. Adélie penguins compensate for the short, severe summers in other ways, which are discussed in chapter 5.

DELAYED RETURN TO THE COLONY

In regard to the effects of age and experience on the annual cycle, several significant facts emerge from the information presented in this chapter: Adélie penguins return to their natal area only after two or more years at sea, females return at younger ages than males, older birds return earlier in the season than younger birds, breeders return earlier than nonbreeders, males return earlier in the spring than females, and most nonbreeders spend fewer days at the colony and leave earlier in the fall than older birds and breeders, respectively. Many of these characteristics have also been reported in studies of other seabird species in which marked individuals have been studied over a long period. Taken together, these patterns constitute a profile common to most migratory, long-lived seabirds, with some variation. For example, male and female black-legged kittiwakes (*Rissa tridactyla*) return at the same time,[20] but the male royal penguin (*Eudyptes chrysolophus schlegeli*) returns slightly earlier than the female.[17] Overall, the age-related patterns of arrival and attendance in the Adélie penguin are remarkably similar to those of the royal penguin, except that royal penguins molt ashore at their colonies. Sedentary species such as the yellow-eyed and gentoo penguins exhibit very few of the these characteristics (see ch. 6).

Why some species spend more years at sea before revisiting the natal area is not clear, but presumably it has something to do with the degree of learning needed to deal with the complexities of existence. Like Adélie penguins, seabirds such as the royal penguin,[16,18] Manx shearwater (*Puffinus puffinus*),[25] Atlantic gannet (*Sula bassana*),[34] short-tailed shearwater (*Puffinus tenuirostris*),[44] and black-legged kittiwake[57] return to natal colonies in significant numbers only after their second or third years. Some others, such as black-footed and Laysan albatrosses (*Diomedea nigripes* and *D. immutabilis*),[41] South Polar skuas,[8] and sooty terns (*Sterna fuscata*)[24] wait until their third or fourth year.

Like Adélie and royal penguins, kittiwakes return during the spring in a temporally segregated pattern:[20] birds with breeding experience come first and are followed in sequence by inexperienced breeders and then nonbreeders. Gannets show a similar pattern: mature males arrive first, followed in order by mature females, young birds in adult plumage, and then immature birds. Furthermore, as noted also for Adélie penguins, young birds spend only brief periods at the colony.[33,34] Like Adélie penguins, Manx shearwaters and sooty terns arrive progressively earlier with age.[24,25] In fact, the visits of young terns are so brief that they merely fly above the colony but never land.

From a proximate standpoint, the age and date of first return for Adélie penguins are related to levels of maturity and annual cycles in reproductive physiology[1] and behavior, including migration.[2] From an ultimate standpoint, maturation rate probably is adaptive and may evolve in response to intraspecific competition for food, space, and breeding partners, as noted for other bird species.[22] This could be so, but to apply the idea to Adélie penguins we must first assess the advantages or disadvantages for the individual that delays visits to a colony or limits first visits to short duration. To do so, we must address two questions: When do the advantages of visiting the colony outweigh those of staying away? Are there advantages to staying away or not visiting?

Penguins need time to perfect their travel and food capture skills.[10] Securing food from the sea is difficult,[11] and many seabird species exhibit greater proficiency with age in capturing prey.[15,21,35,39] For the Adélie penguin, the available information indicates that young birds have more difficulty in obtaining food and that they acquire more skill with practice. Especially telling are the observations that young nonbreeders spend more time away than older nonbreeders between their occupation and reoccupation period visits (see table 5.16), young breeders are away longer than older breeders on foraging trips during incubation (see table 5.16), chicks of older parents reach the size at which they can be left alone at younger ages (with more robust bodies) than those of young parents (ch. 6), and chicks of young breeders fledge at lighter mass than the chicks of older breeders (see table 6.1). These are all fairly direct measures of the foraging and provisioning capabilities of Adélie penguins as a function of age and experience at sea. (The consequences of not learning quickly enough are discussed in ch. 9.)[56]

The fact that experience reduces the degree to which ice conditions affect spring migration indicates that travel efficiency is improved by practice, too. Practice in feeding can take place only at sea, and the long migration to the colony and the period of fasting involved can be accomplished only after successful feeding has built up fat reserves. Being in prime physical condition increases the chances of avoiding predation by leopard seals, which in summer concentrate at the narrow entrances to the colony. Until food capture and swimming skills are perfected, it is advantageous to stay at sea in the pack ice.

Travel to the colony uses the most time and energy early in the breeding season, when sea ice conditions are most severe and when migration may cover hundreds of kilometers. Later in the summer, although significant predation by leopard seals at colonies still occurs, more abundant food and less extensive pack ice mean that penguins can travel much more rapidly. The advantages of visiting the colony then might outweigh the dangers. In fact, young birds of less experience concentrate their visits during the late summer reoccupation period. This is also the time when closely related penguin species (gentoo and chinstrap), which avoid pack ice, begin their annual breeding effort. They visit when conditions are less difficult.

In contrast, the average departure dates of nonbreeding Adélie penguins, especially the youngest ones, correspond fairly closely with the time at which parents discontinue occupation of the colony and leave their chicks in the crèche. When sea ice retreats from the vicinity of the colonies, Adélie penguins find it no longer worthwhile to visit a colony without a specific purpose (e.g., to feed chicks). Young birds therefore cut short their visits and follow failed breeders and older nonbreeders to the distant and retreating pack ice. There they find food more easily and feed heavily to build up fat reserves for the long fast of the annual molt.

The importance of annual and seasonal extent and proximity of sea ice in this species's natural history is a central theme continued in the next two chapters.

Chapter 5

The Occupation Period

Pair Formation, Egg Laying, and Incubation

Chapter 4 described the timing of various phases of the Adélie penguin's annual cycle. The beginning of the nesting cycle—migration to and arrival at the colony—was looked at in detail, as was time spent in the colony and the end of the nesting cycle, the molt. This chapter describes the major events of the occupation period. Included are age of first breeding; territories, nests, and nest building; pair formation; egg laying, eggs, and clutch size; and incubation and hatching success.

■ Age at First Breeding

In a study at Cape Crozier, the modal age of first breeding was five to six years (table 5.1). The average age of first breeding approximated the modal ages: 5.0 years in females and 6.2 years in males ($t = 10.99$, $P < .05$; table 5.2). These averages differ slightly from those used by Ainley and DeMaster[3] (4.7 and 6.8, respectively), who used a different method of calculation. No males bred at three years of age, whereas 6 percent of all female first-breeders were of that age. Assuming an equal sex ratio among eventual breeders (and therefore at first breeding)[3] and correcting for co-

Table 5.1 Age at first breeding by Adélie penguins at Cape Crozier, corrected for mortality and cohort size, 1964–1969.

	A	B	C	D	E	F
Age (yr)	Number Observed	Number Originally Banded	Proportion Surviving*	Corrected Cohort (B × C)	Number of Breeders (A/D × 1,000)	Percentage First Breeding
3	44	21,479	0.20	4,296	10	3.0
4	146	16,420	0.14	2,299	64	19.2
5	98	11,420	0.09	1,028	95	28.4
6	42	6,420	0.07	449	94	28.1
7	7	2,310	0.05	116	60	18.0
8	1	2,308	0.04	92	11	3.3

*Data from Ainley and DeMaster.[3]

Table 5.2 Age at first breeding by female and male Adélie penguins at Cape Crozier, 1964–1969.

	Males			Females		
Age (yr)	Observed n	Corrected* n	Corrected %	Observed n	Corrected* n	Corrected %
3	0	0	0.0	28	6	5.9
4	35	15	3.9	75	33	32.3
5	72	70	18.2	28	27	26.5
6	56	125	32.6	12	27	26.5
7	19	163	42.4	1	9	8.8
8	1	11	2.9	0	0	0.0
Total	183	384	100.0	144	102	100.0
	5.3 ± 0.1	6.2 ± 0.1		4.6 ± 0.1	5.0 ± 0.1	

*Corrected by multiplying the number observed by the ratio obtained by dividing column E by column A from table 5.1.

hort size, 4 percent of first-breeding females were three years old and 7 percent of first-breeding males were four years old. Only 34 percent of females bred first after the fifth year, whereas 60 percent of males bred first at ages six years and older.

Most first-time breeding females visited the colony in one previous

Penguin Problems

Werner Heisenberg may have meant his uncertainty principle to apply only to particle physics. But there is uncertainty in just about all scientific work. That is, the act of measuring or observing something can affect those measurements or observations.

Take studies of penguin behavior. For decades, scientists have been banding the birds, attaching numbered metal or plastic strips to their flippers. That has prompted a debate among researchers; some argue that the bands may affect behavior by increasing drag and making it harder for penguins to swim.

Now, Dr. Michel Gauthier-Clerc of the National Center for Scientific Research in Strasbourg, France, has provided strong evidence that banding does affect the birds. In a five-year study of king penguins on Possession Island in the southern Indian Ocean, Dr. Gauthier-Clerc and his colleagues showed that banded birds often returned to the colony later than unbanded ones, had a lower probability of breeding and produced fewer offspring.

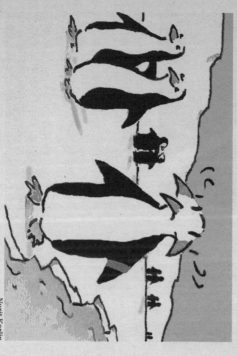

Nurit Karlin

Dr. Gauthier-Clerc, whose study is published in Biology Letters, said it appeared that the impaired swimming affects the birds on critical foraging trips. King penguins normally increase their weight 50 percent on a three-week trip that precedes the November breeding season. "We think the problem is in body reserves," Dr. Gauthier-Clerc said. "If there are not enough reserves, the penguins can't engage in reproduction."

To track his 50 banded and 50 unbanded birds, Dr. Gauthier-Clerc used a newer technology. Tiny electronic transponders were implanted under the birds' skin, and antennas to pick up the identifying signals were placed near the colony.

This technology could easily replace banding in most types of studies, Dr. Gauthier-Clerc said. Although many researchers have abandoned banding for all but short-term studies, a few still practice it for long-term research and are likely to continue doing so, Dr. Gauthier-Clerc acknowledged. "They'll continue to say there is no effect on their penguins," he said. But having shown that bands can affect at least one species, he added, "at least now we can ask, 'Show us the data.'"

A Desire to Be in Control

A second surprise has been the kind of people who use the law. They are not so much depressed as determined, said Linda Ganzini, a professor of psychiatry at Oregon Health Sciences University. She led a recent survey of 35 doctors who had received requests for suicide drugs. The doctors described the patients as "feisty" and "unwavering."

A third lesson is that for most of those who seek assisted suicide, the greatest concern appears not to be fear of pain but fear of losing autonomy, which is cited by 87 percent of the people who have taken their lives with the drugs. Only 22 percent of the patients listed fear of inadequate pain control as an end-of-life concern, perhaps a sign that pain management has improved over the years.

And though opponents of the law argued that patients would feel pressured by families and even insurers to end their lives early out of financial concerns, so far concerns of being a burden to family have been cited by 36 percent of patients, and financial concerns by just 2 percent. The surveys show that the standard version of health care for terminally ill patients might not be what these patients are looking for, Dr. Ganzini said. The standard version of care says, in effect, "we're going to take care of you," she said. But "for them, the real problem is other people taking care of you."

Ms. Jackson said the surveys were changing the hospice association's practices.

"As I helped her into the car, she said, 'He wants me to kill myself,'" Dr. Stevens recalled. "It just devastated her that her doctor, her trusted doctor, subtly suggested that."

Others who initially opposed the law, like the hospice group, say they have learned to live with it. Michael Bailey, for example, took out a loan in 1994 to fight the Death With Dignity act. His daughter has Down syndrome, and he said that at the time he could see a straight line between voluntary assisted suicide and forced euthanasia for the handicapped.

Now Mr. Bailey says he has not seen any abuses. "I don't see that there's ever been a scandal," he said, "and the numbers are not huge." Still, he does not support the law. "If it was up to me, I'd say no, but I don't think there's any great human rights crisis here," he said.

Support for the law crosses ideological lines, said Nicholas van Aelstyn, a lawyer in San Francisco who works with Compassion in Dying. Some commentators have characterized the movement as a liberal cause, but "to most of the people exercising it, it's a libertarian issue," he said. "Many of our clients are die-hard Republicans who don't want government interfering in their lives."

That certainly describes Mr. Wilson, who calls himself a "staunch conservative" and says Mr. Ashcroft is "dead wrong" about the Oregon law.

The support for the law in Oregon, Mr. James said, reflects the pioneer spirit that

"I went into medicine to help people," said suicide. "I didn't go into medicine to give p

educated, family-oriented people willing to hack a new life out of this wilderness," he said. "Pretty independent folks."

Those who drafted the Death With Dignity Act say they did not try to come up with a political document that would warm the heart of Jack Kevorkian, or that would permit euthanasia, which is repugnant to a significant portion of the population. Instead, they say, they carefully drew up a law that they believed would gain support of every-

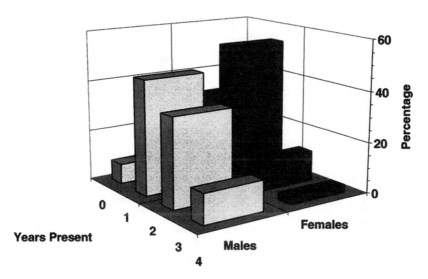

Figure 5.1 The number or years of experience (years present in the colony) of first-time breeding male and female Adélie penguins at Cape Crozier, 1964–1969; birds breeding on their first visit listed as 0 years present. *Data from Ainley et al.*[5]

year. Very few bred on their first visit or remained as nonbreeders for more than two years (fig. 5.1). Males remained nonbreeders but visited the colony for more seasons before breeding for the first time. Fully 86 percent of females were present at Cape Crozier for their first or second season when they first bred, compared with 55 percent of males. The greater delay in breeding for males is related to an unequal sex ratio among older Adélie penguins.[3] The greater number of males increases the competition for a smaller number of available mates, and younger males more often lose in the contest with older males.

The incidence of breeding increased with age (table 5.3). In the study at Cape Crozier, no two-year-olds bred, and only 9 percent of three-year-olds did so. Large increases in the proportion of birds breeding continued successively until the sixth year, and then the trend leveled off. Similar increases in incidence with age occurred in birds with an equal number of years present as nonbreeders in the colony (fig. 5.2A). This was more evident in young birds because with increased age, the effect of added experience decreased. For example, among four-year-olds, overall breeding incidence was 30 percent (table 5.3), but within this group, breeding increased from 10 percent of those present for the first time to 20 percent of those present for their second and 43 percent of birds at the colony for the third season. Among five-year-olds, breeding incidence increased with two or three years of additional experience but not with a fourth, and among six-year-olds only one additional year of experience had an effect. Among older birds, experience did not increase breeding incidence ($P < .05$, t test). In other words, after two years of visiting the colony, additional years of nonbreeding presence did not affect breeding incidence.

Breeding incidence increased more rapidly with prior breeding experience (fig. 5.2B) than with mere presence in the colony. For example, 18 percent of four-year-olds without previous breeding experience bred, whereas 55 percent of four-year-olds that had previously bred once before did so again ($P < .05$, t test). After the first year of breeding, additional experience did not increase breeding incidence. This is revealed in the combined data for five- to seven-year-olds, for which there were no statistically significant differences among birds having one or two years of

Table 5.3 Incidence of breeding in Adélie penguins of known age and sex at Cape Crozier, 1967–1975.

Age (yr)	Males		Females		Total	
	Percentage Breeding	*n*	*Percentage Breeding*	*n*	*Percentage Breeding*	*n*
2	0	104	0	115	0	219
3	0	244	17	228	8	472
4	13	256	49	226	30	482
5	40	449	68	333	52	782
6	66	527	82	414	73	941
7	81	391	90	333	85	724
8	83	198	91	164	87	362
9	79	78	87	46	82	124
10	89	64	89	36	89	100
11	76	59	82	33	78	92
12	82	38	83	24	82	62
13	88	16	100	8	92	24
14	100	4		0	100	4
r^*	0.8509		0.7640		0.8298	

*$P < .05$.

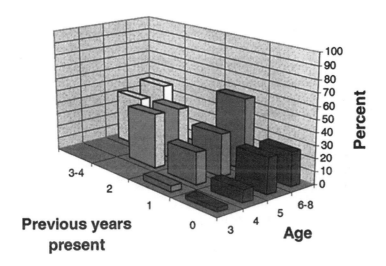

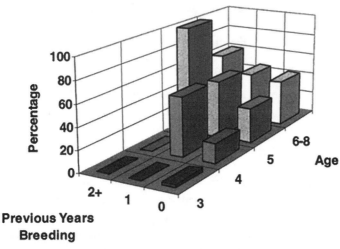

Figure 5.2 Breeding incidence of Adélie penguins as affected by age and number of seasons having visited the colony as nonbreeders, and age and the number of years having bred previously; Cape Crozier, 1964–1969. *Data from Ainley et al.*[5]

prior breeding experience ($P < .05$; t test). There was also no effect of breeding experience on breeding incidence after five years of age, because in six- and seven-year-olds (combined), breeding incidence was the same regardless of experience ($P < .05$). It appears, then, that a year of previous experience speeds the rate of maturation in younger Adélie penguins, at least with regard to increasing their tendency to breed. With few exceptions, incidence of breeding was higher among females than males of the same age, experience in the rookery, and breeding experience.[5]

■ Nests and Territories

An assemblage of nesting Adélie penguins is called a colony. Colonies are further divided into breeding groups or subcolonies, composed only of contiguous territories (ch. 3; fig. 3.2). The subcolonies are located on hummocks, raised ground, or the like; between are gullies or low areas subject to flooding and drifting snow. Adélie penguins use such habitat throughout their range.[26,48,50] The closely related chinstrap penguin tends to use the slopes of steeper terrain, and the gentoo penguin uses flatter terrain close to or on the beach. Nesting on windswept crests of hummocks and raised ground prevents snowdrifts or running meltwater from inundating nests.[24,37,55] These are factors with which the Adélie penguin must commonly contend because of its presence at high latitudes,[12,21,48,50] and they were recognized by the first naturalists to visit Adélie penguin colonies:

> Viewed before the penguins' arrival in the spring, and after the recent winds had swept the last snow falls away, the rookery is seen to be composed of a series of undulations and mounds, or "knolls," while several sheets of ice, varying in size up to some hundreds of yards in length and one hundred yards in width, cover lower lying ground where lakes of thaw water form in the summer. Though doubtless the ridges and knolls of the rookery owe their origin mainly to geological phenomena, their contour has been much added to as, year by year, the penguins have chosen the higher eminences for their nests; because their guano, which thickly covers the higher ground, has protected this from weathering and the denuding effect of the hurricanes which pass over it at certain seasons and tend to carry away the small fragments of ground that have been split up by the frost. (Levick, pp. 9–10)[20]

NEST LOCATION: RELATION TO NATAL SITE

During their first year of visiting the colony, most Adélie penguins wander about and occasionally occupy nest sites for short periods.[5,38] Many young birds wander over a large part of the colony but at some point spend time in the vicinity of where they hatched; they also occupy territories or practice pairing near the natal site (table 5.4). In light of this behavior, not surprisingly, the site where an Adélie penguin first breeds is also related to where it occupied territory as a nonbreeder and where it was hatched (table 5.4). More than half of all Adélie penguins eventually breed within the same subcolony where they hatched and may compete with their parents for nesting space. Slight sexual differences in the importance of the natal site relative to the eventual breeding site are evident. Males spend more time wandering than do females and are thus the primary "prospectors" within the population.[5] As a result, more females breed within the areas and subcolonies of their hatching than do males (both significantly greater; $P < .05$, t test). Similarly, females show a significantly greater tendency to breed in an area of previous wandering than do males ($P < .05$, t test).

NEST LOCATION: EFFECT OF AGE

Young, unestablished breeders tend to switch nest sites after breeding at one site the first year.[19,27,38] This contrasts with the extreme faithfulness to a nest site among older, established breeders. However, the move-

Table 5.4 Percentage of Adélie penguins observed wandering and occupying territories as young nonbreeders in the vicinity of eventual breeding site and percentage of individuals breeding in vicinity of natal site, Cape Crozier, 1964–1969.

Same:	Colony		Area*		Subcolony	
	Percentage	*n*	*Percentage*	*n*	*Percentage*	*n*
Wandering vs. breeding site	97	260	82	271	74	133
Territory vs. breeding site	99	735	98	673	96	596
Breeding vs. natal site	96	337	77	114	56	57

*Area = within 200 m, which is the size of the entire Cape Royds colony.

Table 5.5 Percentage of known-age Adélie penguins nesting at the periphery versus the interior of subcolonies at Cape Crozier, 1968–1974.

	Percentage of Nests by Number of Nests from Periphery of Subcolony					
Age (yr)	0	1	2	3	4+	n
3	38	31	12	6	12	16
4	58	18	13	8	3	77
5	52	27	10	5	6	92
6	56	25	12	4	3	354
7	50	26	16	4	4	283
8	55	29	16	5		59
9+	40	34	15	9	2	162
7–9*	47	29	15	6	3	504
3–5*	54	24	11	6	5	185

*Difference between these categories significant, $P < .05$ (t test).

ment is almost always within the same subcolony. Add to this pattern the fact that arrival time during spring advances with age (ch. 4), and one might surmise that central nest sites are already occupied when younger breeders arrive. As a result, 54 percent of three- to five-year-olds compared to 47 percent of birds seven years and older nest at the peripheries of subcolonies at Cape Crozier ($P < .05$; t test; table 5.5). Three-year-old females were exceptional for their age group because most pair with older males, which tend to occupy more central territories. This tended to dilute the general pattern of younger birds taking peripheral territories.

In addition to taking peripheral nests, young birds also took ones well into the interior of subcolonies. No Adélie penguin older than seven years ($n = 221$) nested more than five nests from a subcolony edge, and 1.4 percent of these birds occupied the fifth nest in. A significantly larger proportion of younger birds (3.8 percent, $n = 822$; $P < .05$, t test) occupied nests that were four or more sites from the periphery. In contrast, disregarding three-year-olds (all females nesting with older birds), more birds that were eight years and older nested within one or two sites of the subcolony periphery than did younger birds: 47.1 percent ($n = 221$) versus 38.2 percent ($n = 822$; $P < .05$), respectively. Older Adélie penguins thus avoided the most central nests, probably because they were too difficult to reach in the process of finding nest stones. Because an Adélie must trespass on every territory along its path into a subcolony, the chances of suffering an outright attack[1,42] instead of a mere peck[1,42] increases with the number of territory boundaries violated.[27]

Imagine an Adélie that has established a central territory and must travel to the outside of the colony to collect hundreds of stones for its nest. No wonder thievery has developed in this species! The frequency of agonistic encounters is high in the center of a subcolony, so it is not surprising that the centrally nesting Adélie penguins tend to be less tolerant of one another (and perhaps more aggressive[39]) than their peripheral counterparts. Therefore, older, more experienced birds tend to avoid nests deep within large subcolonies. It might also be that younger, less experienced birds are more attracted to the increased activity and greater social stimulation in subcolony centers.[53] The subject of central and peripheral territories is discussed further in chapter 6.

TERRITORY AND NEST QUALITY

What does not seem to be affected by age is the size of the territory, if only because not much variation in territory size exists. Within a subcolony, Adélie penguin territories are only slightly bigger than the nests. Barely enough space exists around the nest to allow the nonincubating partner to avoid being pecked by an incubating neighbor; birds on adjacent nests rarely face one another[41] because aggression would ensue. In fact, the spacing of nests is a result of birds on adjacent territories pecking at one another. The loser of the joust shifts location a bit. The result is that nests are spaced with an average 77 centimeters between centers[26] so that a penguin on its nest is just out of reach of birds occupying the four to six territories contiguous with it. The average area per territory measured at a colony in Wilkes Land was 0.75 square meters (range 0.49–0.92 square meters[27]), with variation in spacing related to the proximity to physical features (e.g., rocks, gullies). Hypothetically, assuming a regular, gentle slope and the maximum six contiguous neighbors, territories should be 0.4–0.5 square meters.

The quality of the nest increases with the age and experience of its owners (fig. 5.3). This is more evident among lone males because once they obtain a partner the quality of the nest increases markedly. This is because one bird can defend the nest and its stones while the other is off obtaining more stones. Nest quality increases steadily to age six years

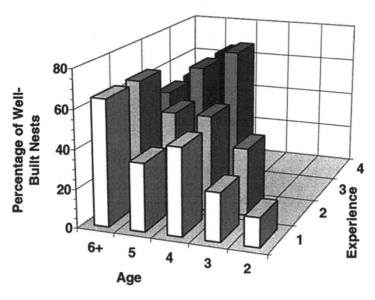

Figure 5.3 Nest quality as a function of age and previous experience in the colony among young Adélie penguins; Cape Crozier, 1964–1969.

among birds present for the first time. However, added experience (years in the colony) has most effect among the youngest birds.

Other factors can affect nest quality. Nests located where there is an abundance of stones are on average more robust than those where stones are scarce. Therefore nest quality must be judged within a subcolony, not between subcolonies. Weather also affects nest size and quality, as Levick notes:

> Owing to our having several snowfalls without wind, and to the action of the sun on the black rock, . . . the rookery became a mass of slush in many places, and in some of the lower-lying parts actually flooded. In some of the low-lying situations penguins had unwarily made their nests, and there was one particular little [sub]colony near our hut which was threatened with total extinction from the accumulation of thaw water. As this trickled down from the higher ground around them, the occupants of the flooded ground exerted all their energies to avert this calamity, and from each nest one of its tenants could be seen making journey after journey for pebbles, which it brought to the one sitting on the nest, who placed stone after stone in position, so that as the water rose the little castle grew higher and higher and kept the eggs dry. One nest in particular

I noticed which was as yet a foot or so clear of the water and on dry ground; but whilst the hen sat on this the cock was working most energetically in anticipation of what was going to happen, and for hours journeyed to and from the nest, each time wading across the little lake to the other side, where he was getting the stones. (pp. 65–66)[20]

■ Pair Formation and Retention

Established breeders perform few displays or vocalizations[2] and instead act directly during pair formation. On the other hand, young, unestablished birds display a great deal as they seek partners and form pair bonds. Individuals known to have bred in a previous season, through their direct approach, establish or reestablish the pair bond (by performing copulation) soon after arrival, females in 1.2 ± 0.1 days and males in 4.5 ± 0.5 days.[5] Birds that have not bred previously take much longer to form pair bonds, females in 4.5 ± 0.2 days and males in 12.3 ± 0.6 days. Age has little bearing on the number of days involved.

The number of days males take to acquire a mate, especially those who have never before bred or whose previous mate did not return, depends on the date of those males' initial arrival (fig. 5.4). Because of the earlier date of arrival (on average) among males and the insufficient number of females

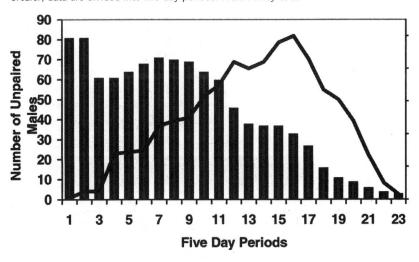

Figure 5.4 The number of unpaired male Adélie penguins (vertical bars) and their proportion (lines) to total males present during the occupation and egg-laying period at Cape Crozier; data are divided into five-day periods. *From Ainley et al.*[5]

present at the beginning of colony occupation, the earliest males usually must wait several days before obtaining a mate. The females that do arrive early mostly are ones that have bred before. Pairing is rapid. Soon thereafter and up to the peak of egg laying, the proportion of unpaired males reaches a lower plateau, indicating that many birds, both males and females, are arriving and are pairing quickly. In contrast, late in the egg-laying period the number of unpaired, territorial males reaches maximum, and many remain unpaired longer. Competition for mates among males, many of whom have not bred before, is intense at that time.

AGE OF MATED PAIRS

In the sedentary yellow-eyed penguin, young females mate predominantly with older males because of the male's attractiveness in terms of dominance, whereas older females mate with younger males because of their greater availability.[33] This pattern for yellow-eyed penguins might also be related to their sedentary nature, which would result in young and old birds being present simultaneously for long intervals well before the egg-laying period begins. For the migratory Adélie penguins, the staggered dates of arrival (ch. 4) and the rapidity of pairing should reduce the chances for old birds and young birds to pair. It would also be advantageous for an old bird to avoid pairing with a young bird (see ch. 6).

Near the end of the long-term project at Cape Crozier, when many known-aged, banded birds were present (1968 to 1975), researchers encountered 103 known-age pairs that produced eggs. To investigate the degree to which birds of various cohorts paired, the number of observed pair combinations was divided by the maximum number of such combinations possible (table 5.6). Results showed that Adélie penguins of a given age tended to pair with others at or near the same age. The pattern exhibited by five- through eight-year-olds was particularly instructive because all age combinations were possible; five-year-olds paired only with birds younger than eight, and eight-year-olds paired only with birds older than five years. Six-year-olds were at a pivotal age and showed the widest array of pair combinations.

WEAK PAIR AND SITE BONDS IN YOUNG BREEDERS

In the yellow-eyed penguin,[33] black-legged kittiwake,[9] South Polar skua,[6,52] and red-billed gull (*Larus novaehollandiae*),[23] about 80 percent

Table 5.6 Weighted* number of pairings for possible age combinations among pairs in which both birds were of known age at Cape Crozier, 1968–1975.

Age (yr)	Age of Mate (yr)								
	3	4	5	6	7	8	9	10	11+
3		**5.4**	**1.1**	1.0	0	0			
4	**5.4**	**2.7**	1.5	0	0	0			
5	1.1	1.5	**2.4**	0.7	**0.8**	0	0	0	0
6	1.0	0	0.7	**2.9**	**1.0**	0.7	0.4	0.4	0.4
7	0	0	0.8	1.0	**3.2**	1.0	**1.5**	0.4	0
8	0	0	0	0.7	**1.0**	0	0.5	**0.6**	0
9		0	0.4	**1.5**	0.5	0	**2.2**	0	
10		0	0.4	0.4	0.6	**2.2**	0	**2.5**	
11+		0	0.4	0	0	0	**2.5**	0	

Data from Ainley et al.,[5] p. 9.

The table is read left to right; a space designates no age combination possible; bold type highlights the two highest numbers of pairings in each row.

*The observed number of pairs for each combination divided by the number of breeding males or females, whichever was less, available for respective age combinations times 100.

of older individuals retained their mates from one season to the next. This percentage was true also for Adélie penguins at the Windmill Islands, Wilkes Land.[27] In contrast, at Cape Crozier only 18 percent of three-year-old breeders who returned the next season bred with the same mate. Among all older age classes of Adélie penguins, 51 percent retained the same mate.[5] In addition, no banded Adélie penguin from Cape Crozier bred with the same mate for more than four years. In fact, in a sample of 100 pairs of banded birds, only six penguins bred three seasons with one mate; among twenty-two birds at Crozier that bred for four or more seasons, one-third bred with four different mates, and half bred with three different mates.[5]

Analysis of cases at Cape Crozier in which known-age Adélie penguins and mates returned and the known-age bird bred (table 5.7) indicates marked differences in mate and site tenacity between young, unestablished breeders and presumably older, established breeders. In this situation, only 56 percent of young breeders compared to 84 percent of established adults retained their previous mate ($P < .05$). Fifty percent of

Table 5.7 Mate and site fidelity among Adélie penguins when both members of a pair returned and the known-age partner bred, Cape Crozier, 1965–1968.

	Known-Age Penguin (yr)								Total		Adult
	7		6		5		4				
	F	M	F	M	F	M	F	M	n	%	%
Retained mate	3	1	4	2	9	2	2	0	23	43	84
Disunited	2	2	10	2	9	0	5	1	31	57	16
Retained site	3	3	6	1	9	2	1	0	25	49	78
Moved >1 meter	2	0	7	3	8	0	6	0	26	51	22

Adult data from Penney.[27]
Adult–young differences significant ($P < .05$, t test).

twenty-six young breeders who did not pair with their former mate retained their former site within 1 meter (versus 78 percent in adults); the remainder moved farther than 1 meter.

This contrast between unestablished and established breeders is evident from the history of one subcolony of about 150 breeding pairs at Cape Crozier (fig. 5.5). In five breeding seasons, three young females bred with eight (possibly nine) different mates. One of these paired with two different young birds in consecutive seasons. Only once did a pair remain united for two seasons, even though in all but two instances the mate from the previous year was present and available for breeding. In the sample of ten presumably established pairs banded at Signy Island in 1946, at least six were together the next year, and only two mated with other birds when the previous mate was available.[38] Five stayed together for three seasons.

Asynchrony of return to the colony is a primary factor in the disuniting of pairs.[5,27] In about half of the disunited pairs at the Windmill Islands, return was widely asynchronous, with seven males and eight females returning seven or more days after the partner's arrival. Younger birds that disunited from mates had even greater asynchrony in arrival at Cape Crozier: Partners in eight of ten pairs that disunited arrived seven or more days apart, and in five pairs arrival was separated by ten or more days. However, asynchrony of arrival does not always lead to switching of mates. Two of the cases at Cape Crozier in which known-age birds retained mates (table 5.7) involved males arriving as much as eighteen days before their mates. In all cases when pairs reunited, the male arrived first.

Plate 1 Adélie penguins in their fishlike phase, passing beneath the water surface as darting shadows.

Plate 2 The transition from land dweller to sea creature occurs in the blink of an eye, after much mental preparation.

Plate 3 Adélie penguins burst from the sea, eyes wide, exuding an unparalleled energy that they must subdue before tackling colony life.

Plate 4 Most penguins in the world simply walk ashore, as these Adélie penguins are doing.

Plate 5 An Adélie penguin colony showing the groups, called subcolonies, from a much closer perspective than that shown in figure 3.2.

Plate 6 Two Adélie penguins in the initial stages of pair formation and breeding: suspicious male on the left, not-yet-convinced female on the right.

Plate 7 An Adélie penguin parent viewing the contents of the egg it had been sitting on for many long days.

Plate 8 A small chick receives its first meal; day by day it will grow in size from a few to several hundred grams.

Plate 9 A chick too large to be brooded entirely by its guarding parent; soon it will become too much for the parent and will enter the crèche.

Plate 10 A lone parent being inspected by hopeful, near-to-fledging chicks looking for one last meal before going off on their own.

Plate 11 Unlike most other penguin species, Adélie penguins usually must leap over hurdles to reach land; the one falling back after hitting the wall is at risk of being captured by a leopard seal.

Plate 12 A hungry leopard seal doesn't stand much of a chance in this situation.

Plate 13 A leopard seal captures an Adélie penguin, having surprised it along the edge of an ice floe.

Plate 14 These Adélie penguins are nesting above wave-eroded boulders, vestiges of a beach that has long since become dry because of sea level change.

Plate 15 Adélie penguins nest on a layered deposit of guano 7,000 years in the making.

Plate 16 Adélie penguins surrounded by the vastness of Antarctica and its frozen ocean.

The opposite was true in cases of disuniting. This agrees with information presented earlier: Females seldom remain alone on a nest but almost always pair within a day of arrival at the colony; in contrast, males often occupy territory alone for some time before pairing. Whether the two early-arriving males spent two weeks alone on their nests out of attach-

Figure 5.5 A comparison of typical mate-retention histories between low-latitude Signy Island, 1946–1952,[38] and high-latitude Cape Crozier, 1964-1969. UB = unbanded bird. *From Ainley et al.*[5]

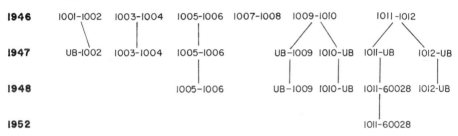
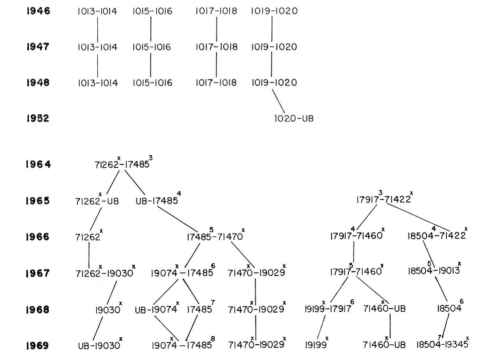

ment to a previous mate or out of inability to secure another mate is open to speculation.

With such low fidelity, it is not surprising that changing mates has little effect on the reproductive success of Adélie penguins at Cape Crozier (ch. 6). The shortness of the breeding season and the consequent importance of arrival and breeding cycle synchrony are no doubt the reason why mate fidelity is so low in Adélie penguins of the southern Ross Sea compared with birds of lower latitudes and with other long-lived seabirds.

■ Egg-Laying Date

GEOGRAPHIC AND TEMPORAL PATTERNS

Not surprisingly, the relationship between date of first egg in a respective colony and that colony's latitude is similar to that between colony arrival date and latitude (ch. 4). The date of first eggs in the northernmost colony (Signy Island) and southernmost (Ross Island) differs by about ten days (fig. 5.6). As is true with arrival date, the retardation of laying date climbs steadily with increased latitude from 60° south to

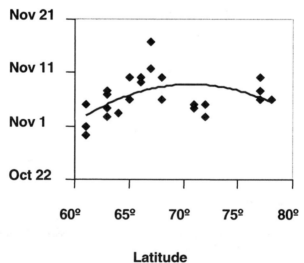

Figure 5.6 The date when the first Adélie penguin eggs were observed at various colonies, by latitude (degrees south). *Data from various sources.*[4,17,21,27,29,37,43,50]

Table 5.8 Mean date of clutch initiation of Adélie penguins nesting in different locations.

Place	Year	Date in November ± SE	Reference
Signy Island	1980	3.0 ± 0.4	21
	1981	3.0 ± 0.3	
Ross Island			
Cape Crozier	1967	11.4	18
Cape Bird	1977	12.6 ± 0.2	11
Lützow-Holm Bay	1990	18.5 ± 0.2	49

about the Antarctic Circle (66.5°S), at which point the relationship reaches a plateau. Thereafter, increased latitude leads to no further retardation in laying. Like that of arrival, this pattern also seems likely to be related to the very rapid change in photoperiod in spring at localities south of the Antarctic Circle. As with arrival date, the colonies whose dates diverge from the relationship are Béchervaise Island (unusually late laying for its latitude) and capes Adare and Hallett (unusually early laying for their latitudes). These anomalies probably are related to access as dictated by sea ice conditions: extensive fast ice at Béchervaise but an adjacent polynya at capes Adare and Hallett.

There are surprisingly few data on the average date of laying for this species (table 5.8). Why this is so is not clear. However, existing data are consistent with the dates of first eggs seen in colonies. Average date of clutch initiation is about eight days earlier at Signy Island (60°S) than at Cape Crozier (77°S). Where the colony is locked in by fast ice, Lützow-Holm Bay (69°S), clutch initiation is delayed additional days (in this case seven). The very limited dispersion of dates around the mean clutch initiation date (two-thirds within about three days) bespeaks the extremely synchronous egg laying in this species.

Egg laying is very synchronous in many different penguin species, including the Adélie.[33] Table 5.9 indicates a high degree of synchrony at Cape Crozier. Half of all eggs at Crozier usually were laid in a six-day period, approximately November 15 to 21. The widest range of egg production within one season (1967) was November 3 to December 2, or twenty-nine days. During 1968 the range was only twenty-three days, from November 4 to 27.

Table 5.9 Mean date (in November) on which the first egg of a clutch is laid relative to the age of the male or female member of the pair, Cape Crozier, 1963–1974.

	Male		Female		Total*	
Age (yr)	Date ± SE	n	Date ± SE	n	Date ± SE	n
3			20 ± 0.9	14	20 ± 0.9	14
4	18 ± 1.0	18	20 ± 0.4	42	19 ± 1.0	60
5	19 ± 0.5	107	18 ± 0.4	116	18 ± 0.3	223
6	18 ± 0.3	310	18 ± 0.3	281	18 ± 0.2	591
7	18 ± 0.3	284	18 ± 0.3	266	18 ± 0.2	550
8	18 ± 0.4	152	17 ± 0.5	134	18 ± 0.2	286
9	18 ± 0.7	47	17 ± 0.7	38	18 ± 0.4	85
10	17 ± 0.8	47	17 ± 0.7	30	17 ± 0.5	77
11	15 ± 0.7	32	17 ± 0.7	20	16 ± 0.5	52
12–14	17 ± 0.6	34	18 ± 0.7	24	17 ± 0.4	58
Total	18 ± 0.2	1,031	18 ± 0.2	965	18 ± 0.1	1,996

*In a regression of laying date versus age, $r = -0.8837$ ($P < .05$).

EFFECT OF AGE AND EXPERIENCE

The date on which an Adélie penguin arrives at the colony and the interval between pairing and laying of the first egg together determine the actual laying date. The latter turns out to be slightly later in November on average for young females and also for the mates of young males (who are also young, as discussed earlier). Apparently, the male yellow-eyed penguin does not affect the ovulation date of his mate;[33] this is probably so for Adélie penguins, too.

Prior breeding experience had an insignificant effect on laying date when age factors were eliminated.[5] On average, pairs containing a bird who was four, six, or seven years old and who had previous breeding experience laid the first egg earlier than pairs with an inexperienced bird of the same age. In the sample of five-year-olds, however, pairs with an experienced bird laid their first eggs later than pairs with an inexperienced five-year-old. None of these differences was statistically significant ($P < .05$, t test), which, with the analysis in table 5.9, indicates that laying date is more a function of physiological condition or maturity than of experience.

It is evident from this and earlier discussions that within several days, the laying date of an Adélie penguin pair is related to the age of the female. Richdale[33] reported a similar relationship in yellow-eyed penguins but also concluded (p. 21) that "individual mean differences in laying dates . . . have a genetic basis"; some females lay consistently early and others consistently late compared with the average laying date of the population. This has been observed among Adélie penguins at Cape Bird[40] as well as short-tailed shearwaters[37] and Manx shearwaters.[15]

Surprisingly, no such relationship was evident at Cape Crozier in eighteen females whose laying dates were known for three or more years.[5] Some females did appear to lay earlier or later, but differences were not statistically significant. The difference between the Cape Bird and Cape Crozier data, other than a slight one in sample size ($n = 28$ vs. $n = 18$), respectively), was that all Cape Crozier females were similar in age and experience. Records began with the first year of breeding in almost all cases, and in that year most ($n = 13$ of 18) were five or six years old, whereas the ages and previous experience of Cape Bird Adélie penguins were unknown. Spurr[40] could have been comparing groups of birds of different age and experience; those younger would lay later and those older would lay earlier than the average for the population. The question of whether some Adélie penguins lay consistently earlier or some later remains unresolved; a study of females of known history for four or more seasons is needed. Because of the short season and the high degree of breeding syn-

Illustration 5.1 An Adélie penguin nest with its complement of two eggs. *Drawing by Lucia deLeiris © 1988.*

chrony in Adélie penguins at Ross Island, however, this phenomenon might not become apparent except in studies at colonies located in lower latitudes.

■ Eggs

CLUTCH SIZE

In most cases, a female Adélie penguin lays two eggs to complete the clutch. The low variance in mean clutch size, both geographically (from colony to colony and region to region) and temporally (year to year), is remarkable (table 5.10). Information on mean clutch size is available for eighteen seasons among seven breeding colonies. During seventeen of those seasons, average clutch size ranged between 1.8 and 2.0 eggs per nest, with 1.9 eggs per clutch being the most common ($n = 9$ seasons). Only in 1968 at Cape Bird, Ross Island, has clutch size been observed to be significantly lower than average (1.6 eggs per nest).[40] Laying at Cape Bird in that year was unusually late because of late arrival of breeding birds, in turn because of extensive sea ice. Therefore the migration was overly difficult, and body resources of females probably were diverted from egg formation to self-maintenance.

Normally, the maximum number of eggs laid by a female in a given season is two. In a two-egg clutch, the second egg is laid two to four days after the first (average 3.0–3.1 days).[7,30,38,40,44] If the first egg is lost before the second is laid, often a third is laid to bring the total to two eggs. Third eggs are laid two to five days after the second is laid (average 3.3–3.4 days).[7,40] No other replacement eggs are laid. Regardless, egg replacement is rare.[38] Among sixty-two clutches observed at Signy Island in 1980, in only 3 percent was a third egg laid; in 1981 among seventy-nine clutches, none received a third egg.[21] At Cape Bird in 1977, a third egg was laid in 1 percent of 452 clutches.[11] In an experimental study at Cape Royds, first eggs were removed from nineteen nests and a third egg was laid in twelve (65 percent).[44] Therefore, in contrast to the rarity of the phenomenon, this species seems quite capable of laying one replacement egg.

Within a given colony, the proportion of nests in which just one egg is laid is about 19–21 percent.[40,44] In 1968 at Cape Bird, when laying was delayed well beyond the expected average date, more one-egg clutches than usual were laid.[40] Therefore the delay in breeding probably was re-

Table 5.10 Summary of nesting success of Adélie penguins by locality and year.

Location	Year	Mean Clutch	Percentage Hatched	Percentage Fledged	Percentage Success	Chicks per Pair	Reference
Ross Is	1959	1.8	66	75	50	0.9	44
Cape Royds	1962		37	72	26		54
	1964		88	77	68		54
	1965		63	90	57		54
Cape Bird	1967	1.9	80	83	66	1.2	40
	1968	1.6	57	76	43	0.7	40
	1969	1.9	72	76	55	1.0	40
	1970	1.8	80	80	64	1.2	40
	1977	1.9	56	63	36	0.7	11
Cape Crozier	1968	2.0	67	96	65	1.3	5
	1969	1.9	75	70	53	1.1	5
Cape Hallett	1960	1.9	82	82	67	1.2	31
King George Is	1977	1.8	64	83	53	1.0	46
Admiralty Bay	1981	1.9	72	73	53	1.0	46
	1982	1.9	53	68	36	0.7	46
	1984	1.9	70	84	59	1.1	46
	1985	1.9	63	81	51	1.0	46
	1986	2.0	75	79	61	1.2	46
Signy Island	1980	2.0	82	47	38	0.8	21
Gorlay Peninsula	1981	1.9	81	94	47	1.3	21
	1982				50		10
	1983				40		10
	1984				40		10
	1985				60		10
	1986				51		10
Magnetic Island	1981					1.2	51
	1982					1.6	51
	1983					1.2	51
	1984					1.0	51
	1985					0.8	51
	1986					1.4	51
	1987					1.5	51
Béchervaise Is	1990					0.8	17
	1991					0.7	16
	1992					0.8	16
	1993					1.1	16

Table 5.10 Continued

Location	Year	Mean Clutch	Percentage Hatched	Percentage Fledged	Percentage Success	Chicks per Pair	Reference
	1994					0.1	16
	1995					0.4	16
	1996					0.9	16
	1997					0.9	16
	1998					0.4	16
Lützow-Holm Bay	1990	1.9	69	70	48	0.9	49
Mean ± SE		1.9 ± 0.1	69.1 ± 2.6	77.1 ± 2.3	51.4 ± 2.2	1.0* ± 0.1	

*Without Magnetic Island data, 0.9.

Table 5.11 Mean clutch size among Adélie penguins of known age, Cape Crozier, 1963–1975.

	Male		Female		Total*	
Age (yr)	Mean ± SE	n	Mean ± SE	n	Mean ± SE	n
3			1.5 ± 0.1	31	1.5 ± 0.1	31
4	1.7 ± 0.1	26	1.8 ± 0.1	92	1.8 ± 0.1	136
5	1.8 ± 0.1	138	1.8 ± 0.1	170	1.8 ± 0.1	349
6	1.8 ± 0.1	321	1.8 ± 0.1	309	1.8 ± 0.1	653
7	1.9 ± 0.1	304	1.9 ± 0.1	287	1.9 ± 0.1	603
8	1.8 ± 0.1	162	1.9 ± 0.1	147	1.8 ± 0.1	313
9	1.8 ± 0.1	52	1.9 ± 0.1	39	1.8 ± 0.1	92
10	2.0 ± 0.1	50	1.8 ± 0.1	31	1.9 ± 0.1	86
11	2.0 ± 0.1	40	1.9 ± 0.1	25	1.9 ± 0.1	66
12	1.8 ± 0.1	31	2.0 ± 0.1	20	1.9 ± 0.1	51
13	1.8 ± 0.1	13	1.5 ± 0.1	8	1.7 ± 0.1	21
14	2.0	2			2.0	2
Total	1.8 ± 0.1	1,139	1.8 ± 0.1	1,159	1.8 ± 0.1	2,403

*Includes data from individuals of unknown sex.

lated to lower-than-usual fat reserves. Late-laying pairs and pairs nesting at the edge of subcolonies are more likely to produce a single egg.[38,40] Such nests are owned by young birds.

Mean clutch size was slightly lower in pairs having a four-year-old male or, in particular, a three-year-old female, where it was 1.7 and 1.5 eggs, respectively (table 5.11). Among older breeders there was no difference in clutch size, which for them averaged 1.8 eggs. If only birds of known history were considered, the only truly different clutch size was that of three-year-old females.[5] Thus breeding experience apparently does not affect clutch size.

EGG SIZE

The size of Adélie penguin eggs varies according to the order of laying, does not appear to vary from year to year, but may vary by locality. Freshly laid eggs weigh 31 to 185 grams but average 101 to 124 grams depending on several factors (table 5.12).[21,25,30,44,54] Whether an egg as small as 31–58 grams is viable is not known.[30] In my experience, very small eggs (half the size of the typical egg), which I have not taken the time to measure, are not viable. These eggs are very rare. The viability of extremely large eggs is not known. In fact, the largest egg in the series reported earlier (185 grams) did not hatch and was found deserted.[44]

Most studies in which eggs have been measured show that the second egg of a two-egg clutch is 7–8 percent smaller and the third egg is 19 percent smaller (by mass) than the first (table 5.12). Eggs laid in single-egg clutches usually are smaller (6–9 percent) than first eggs in two-egg clutches. Single eggs appear to be the most variable in size.[54] The small size of single eggs probably is related to the fact that usually they are laid by young, first-time breeders.[38,44] It is thought that such birds have not quite reached the foraging proficiency of established breeders. Thus young females would have a lesser reserve of body resources from which to form the eggs. Although egg size as a function of female age has not been studied directly in the Adélie penguin, studies of other species, such as the yellow-eyed penguin, have shown that youngest females lay smaller eggs.[33]

If young females lay smaller eggs because they have a lesser ability to attain breeding condition, then it also might be true that more difficult and longer prebreeding migration would result in smaller eggs. Under such conditions, body reserves accumulated before the migration began would have to be diverted from the reserve used to form eggs. Such conditions do reduce the ability of incubating birds to last until return of the mate. However, although the extent of fast ice and concentration of pack ice varied between the four seasons in which egg size was measured at Cape Royds, there was no significant interannual variation in egg size (table 5.12). On the other hand, pack-ice concentration does not vary much at Cape Royds in November compared with, for example, Cape Crozier on the open-ocean side of Ross Island.

Table 5.12 Average size of Adélie penguin eggs.

Location	Year	Number	Mass (g, mean ± SE)	Volume (cc, mean ± SE)	Length (mm, mean ± SE)	Width (mm, mean ± SE)
Single Eggs						
Cape Royds	1959	10	117		69.2	54.1
	1961	15			69.9 ± 0.8	54.6 ± 0.4
	1964	12			69.3 ± 1.0	53.6 ± 0.9
	1965	20	106.3 ± 2.3		68.7 ± 0.4	54.3 ± 0.4
Signy Island	1981	15	111 ± 2.7	99.2 ± 1.9	67.3 ± 0.8	54.0 ± 0.4
First Eggs						
Cape Royds	1959	32	124		69.8	55.8
	1961	123	117	107.9 ± 0.9	69.7 ± 0.3	54.9 ± 0.2
	1964	86		111.2 ± 1.3	70.4 ± 0.3	55.0 ± 0.2
	1965	70		111.2 ± 1.3	70.2 ± 0.2	54.5 ± 0.3
Cape Crozier	1981	79	122.9 ± 1.1		69.8 ± 0.3	55.7 ± 0.2
Cape Hallett	1959	15	124		70.3	56.2
Signy Island	1981	73	121 ± 1.1	106.1 ± 1.0	69.2 ± 0.3	55.3 ± 0.2
Second Eggs						
Cape Royds	1959	21	115		68.7	54.5
	1961	109	104.3 ± 0.9		69.3 ± 0.3	54.2 ± 0.2
	1964	86	106.2 ± 1.0		68.7 ± 0.3	54.4 ± 0.2
	1965	70	108.5 ± 1.2		69.6 ± 0.3	54.5 ± 0.2
Cape Crozier	1981	53	114.8 ± 1.7		68.7 ± 0.4	55.0 ± 0.5
Cape Hallett	1959	15	118		68.9	55.3
Signy Island	1981	73	113 ± 1.1	100.4 ± 1.0	68.4 ± 0.3	54.2 ± 0.2
Third Eggs						
Cape Royds	1959	12	101		66.1	51.4
	1961	8			67.1 ± 1.2	51.2 ± 0.9
	1964	3		80.9 ± 1.8	64.6 ± 0.5	48.7 ± 0.6
Cape Crozier	1981	12	109.1 ± 1.1		67.4 ± 0.7	53.3 ± 0.7

Data from several sources.[7,21,30,43,53]

The eggs measured at Signy Island (60°S) in 1981 were noticeably smaller than those measured over the four seasons, 1959–1964, at Ross Island (77°S) and a few seasons at Cape Hallett (75°S; table 5.12). This is best seen in measurements of the length and width rather than the mass and volume. The latter measurements are more variable and subject to greater interobserver variability. Whether the difference in measurements observed is related to geographic variation or between-year variability is not known. However, in two years of study at Signy, one would expect smaller eggs to have occurred in 1980, the year when eggs were not measured, than in 1981, when they were.[21] In 1980, the sea ice remained in place far longer than usual and well into the nesting season. Thus it is more likely that ice conditions in 1980 forced females to use more of their reserves for migration than for egg formation.

Why there might be geographic variation in egg size in this species is open to conjecture. In general, and in comparison with other avian species, Adélie penguins lay an egg that for its dimensions has a large volume.[30] A larger egg in the colder southern latitudes would retain heat more efficiently, and this pattern would carry over to explain the geographic variation in egg size among Adélie penguins.[30,54] Prey resources might well be more available to Adélie penguins in the Ross Sea than in the waters around Signy Island, and this might explain the difference in egg size as well. Further investigation into the geographic variation in the size of Adélie penguin eggs might be fruitful.

The mass of an Adélie penguin egg is about 2.6 percent of a female's body mass.[30] This proportion is similar to that in other pygoscelid penguins (chinstrap, gentoo) but higher than that in other penguin species (1.6–2.2 percent). The ratio of egg mass to female body mass places the Adélie penguin egg within the group of birds whose chicks are classified as being altricial (that is, immature and helpless) upon hatching. Conversely, precocial chicks (e.g., ducks, shorebirds) are so well developed upon hatching that they can run about soon thereafter. Species whose chicks are precocial lay eggs that are larger relative to a female's body size (about 4 percent larger).

EGG COMPOSITION AND FORMATION

The typical Adélie penguin egg, by mass, is composed of 13 percent shell, 65 percent albumen, and 22 percent yolk.[7,30] Although eggs within a clutch vary in size depending on whether the egg is first, second, or third laid, the size of the yolk does not vary. Therefore it is the amount of albumen that differs among the eggs of a clutch.[7]

The heavy shell of the Adélie penguin egg is typical of avian species that produce precocial chicks, but the proportion of yolk is midway between those of altricial and precocial species.[30,34] According to Reid,[30] the down of an Adélie penguin hatchling is comparable in quality to that of most precocial species, but the chicks' helplessness and poor thermoregulation at hatching[13] shift it toward the altricial category. However, the Adélie penguin hatchling is more advanced (e.g., it is completely feathered) and has a larger yolk sac than many altricial species. Therefore it is better equipped than the initial amount of yolk indicates. The large yolk sac, deriving from the large amount of yolk in the egg, is beneficial to the hatchling, as discussed more fully later in this chapter.

An Adélie penguin female produces each egg in 19–24 days (average 20.7 days) regardless of the egg's position in the clutch.[7] Most of this period, averaging 15.0 days (range 14–17 days), is involved in yolk formation. The yolk is laid down in rings, with each ring representing one day (i.e., fifteen rings). The total amount of time for yolk formation in the Adélie penguin is comparable to that in the larger diving birds of the Northern Hemisphere (e.g., murres [*Uria* spp.] and puffins [*Fratercula* spp.]).[35] Within most avian families, smaller species take less time to form yolks than larger ones, but this pattern does not necessarily hold true when comparing species across families. Not much is known about the rate of yolk formation in other penguins. The rate in Adélie penguins appears to be similar to that in the Fiordland crested penguin (equivalent in size to the Adélie) and slightly longer than in the little penguin.[14]

For each female Adélie penguin, yolks are initiated sequentially about three days apart: 3.0 ± 0.1 days between first and second, 3.3 ± 0.1 days between second and third.[7] These intervals are equal to those between the laying of successive eggs in a clutch. Usually the third yolk is absorbed, a process that is initiated quickly after the second egg is ovulated or laid. In fact, the yolks of third-laid eggs show signs that the absorption process was interrupted by the need to lay the third egg (i.e., the first egg was lost). The interval between yolk completion and laying averages 5.7 days. During the lag, the albumen is added, and in the last day the shell is added. This lag before laying, in which the albumen and shell are contributed, is long compared with that of other avian species and may enable the penguin to reduce its daily protein demand for egg production. The protein needed for egg formation in the Adélie penguin comes from muscle tissue because the penguin is fasting during the process (because of the long migration over concentrated sea ice).

The interval between the female's arrival at the colony and the date she lays her first egg averages 8.4 ± 0.2 days at Cape Crozier, Ross Island.[5] This interval is slightly longer in young females and slightly shorter in older ones (table 5.13). A female must be present for a minimum of seven days to lay fertile eggs.[2] What this short amount of time indicates is that yolk formation is initiated as much as a week before the female arrives at the colony. Fertilization of the egg occurs before the albumen is added. Thus fertilization occurs within the first few days of a female's presence at the colony. This is why a female does not wait long for her mate of the previous year if he is not present when she arrives. Females pair within one to two days of arrival. In most other bird species, egg formation is initiated after the female arrives at the colony or breeding grounds.[35] If that were the case for Adélie penguins, females could wait longer for mates to return. But Adélie penguins cannot afford to wait because of the short summer season.

Table 5.13 Average number of days between arrival and egg laying among known-age males and females that have or have not bred previously, Cape Crozier, 1967–1974.

	Males				Females			
	Prior Breeders		History Unknown*		Prior Breeders		History Unknown*	
Age (yr)	Days (mean ± SE)	n	Days (mean ± SE)	n	Days (mean ± SE)	n	Days (mean ± SE)	n
4			9.7 ± 0.6	18	8.6 ± 1.0	8	9.3 ± 1.5	31
5	12.3 ± 1.1	6	12.4 ± 0.6	101	10.2 ± 0.6	22	7.7 ± 0.4	93
6	13.0 ± 1.0	21	12.0 ± 0.5	172	9.4 ± 0.6	30	7.0 ± 0.3	140
7	12.9 ± 1.1	15	13.5 ± 0.5	144	9.4 ± 0.6	20	7.2 ± 0.3	128
8	13.9 ± 1.3	8	13.6 ± 0.9	36	7.2 ± 0.6	34		
9	16.0 ± 2.1	19			5.3 ± 0.8	18		
10	14.1 ± 1.0	27			5.9 ± 0.5	15		
11	10.8 ± 1.4	18			7.5 ± 1.3	6		
12	16.9 ± 1.9	10			6.7 ± 0.5	9		
13	11.3 ± 2.1	3						
Total	13.1 ± 0.4	127	12.7 ± 0.3	472	8.4 ± 0.2	162	7.3 ± 0.2	392

Statistical comparisons (*t* tests): males versus females, prior breeders and history unknown ($P < .05$); prior breeders versus history unknown, males ($P > .05$) and females ($P < .05$).
*Includes many first-time breeders.

Table 5.14 Incubation period, in days, of Adélie penguin eggs in relation to order of laying.

	Year	First Egg	Second Egg	Single Egg	Reference
Ross Island					
Cape Royds	1959	34.8	33.3	33.3	44
Cape Bird	1969	34.7	33.2	33.8	40
	1970	35.1	33.1	35.3	
	1977	35.2	33.7	34.5	11
Cape Hallett	1959–1962	35.5	34.1		30
Signy Island	1980	35.6	33.5		21
	1981	35.3	33.9		

The first rings to be laid down in the Adélie penguin egg yolk are intensely orange-colored and easy to distinguish.[7] The last-laid rings are a dull yellow and difficult to distinguish. Lee Astheimer and Dick Grau reasoned that the intense colors came from euphausiids, the pink color of which is derived from carotenoid pigments. In most of the eggs that these authors inspected in 1981, about seven to ten of the innermost layers of yolk were stained orange; the remainder of rings were pale. The transition from stained to pale layers indicates the day on which the females start to depend on body reserves rather than on stomach contents to form the eggs. In colonies located farther from polynyas than is Cape Crozier, say Béchervaise Island or Cape Royds (see ch. 3), egg yolks probably are paler. In those circumstances the longer migration over sea ice would mean that the penguins have to depend on body reserves entirely or almost entirely to form eggs.

It is also possible that the early rings are so distinct from the later ones because at the start of yolk formation, the females are far enough north that distinct day-night cycles exist. A day-night difference in yolk deposition is what creates the ring structure of a yolk.[14] Once the females migrate far enough south to be in twenty-four-hour sun, there may not be as much of a day-night cycle in yolk deposition. Therefore the ring structure of the yolks would become more difficult to detect.

■ Incubation

The incubation period of an Adélie penguin egg ranges from thirty to thirty-nine days, depending on the order of laying and other factors. The first egg is not incubated closely (only 47–70 percent of the time) until the second egg of a clutch is laid.[21,30,38,40,44] This does not mean that the first egg is exposed to the elements; it simply means that the birds do not sit tightly. The result is that the two eggs hatch an average of 1.4 days apart (compared to the 3.0 days that separate their laying). The average incubation period is longer for single and first eggs than for second eggs (table 5.14). This means that birds having a single egg do not incubate as efficiently as those having two eggs, a pattern observed in other seabirds.[28]

The Adélie penguin's incubation period is a few days shorter than those of other pygoscelid penguins (gentoo, chinstrap).[30] The shorter incubation period would be an advantage for a species that nests at higher latitude (with a resulting shorter summer). Actually, the short incubation period could be the result of Adélie penguins incubating more tightly, perhaps to protect against the colder temperatures they experience compared with the other pygoscelid species. In fact, the incubation periods of eggs at the most southern (colder) latitudes (Ross Island) are slightly shorter than those of eggs laid by Adélie penguins farther north. However, more studies at more localities are needed to confirm this. On the other hand, the incubation period of Adélie penguin eggs (and those of

Illustration 5.2   A nest relief among a pair of mated Adélie penguins. *Drawing by Lucia deLeiris © 1988.*

most other penguin species) is longer than for eggs of many other species laying eggs of comparable volume.[30] This probably is related to incubation temperature and the fact that penguin eggs, especially those of the polar species, experience short periods of chilling.

NORMAL INCUBATION ROUTINE

Normally, females leave for the sea to replenish their food reserves as soon as they lay the second (or third) egg, completing the clutch. The male continues his fast, begun with migration to the colony and subsequent territory establishment and pair formation. The length of the male's first incubation shift varies widely, from just a few days to the entire incubation of the clutch (and subsequent failure because if he has to remain that long, it is because the female has disappeared). However, for most birds in a given colony the typical routine is as follows, although there appears to be some geographic or intercolony variation. In most cases on Ross Island, it appears that the male's first shift is longer than the female's: eight to fifteen days for the males (usually more than twelve) and two to sixteen days for the female (usually fewer than twelve; table 5.15). The male then takes another turn that lasts five to six days, followed by several trades between mates for one or two days until the eggs hatch. Thus the mates typically switch eight or nine times. In contrast, at Signy Island (and Cape Hallett), the first stints of the male and female are about equal, with both being thirteen to fifteen days. There is then a five- to eight-day turn by the male, with the female returning again at about the time of hatching. In other words, there are four nest reliefs. The scheme at King George Island has elements of both the Ross Island and Signy Island schemes and possibly is more often like the scheme at Signy Island.

The scheme at Ross Island seems well suited to a system in which the rules are dictated by abundant food and by extensive, compacted sea ice through the period of egg laying, with a high likelihood that the ice will loosen soon thereafter. The extensive, compacted sea ice increases the difficulty of foraging. Such a system works well where a polynya adjacent to the colony is an annual certainty. In 1962 at Cape Royds, open water occurred right to the beach during the egg laying and incubation periods. Mates returned so soon that the incubating birds often did not want to relinquish incubating duties.[44] On the other hand, the scheme at Signy Island (and perhaps Cape Hallett) seems well suited for a system in which there is not much sea ice, but food availability, independent of sea ice, may be problematic. In other words, the birds need all the time possible to search for food and forage sufficiently to restore their body condition. The scheme at King George Island appears to vary as a function of food availability. In 1982 it appeared to be similar to the general condition at Signy. That year at King George was deemed to be one of low food availability.[46] In 1985, the King George scheme was closest to the Ross Island scheme; in that year the highest breeding success in the several-year study was seen (therefore indicating abundant food). The King George Island data thus seem to support the hypothesis that the extremes of incubation routine schemes (Ross and Signy) are related to food availability.

Normally, the male incubates on average 61 percent of the days needed to hatch the eggs. However, the proportion varies by age (and sex).[5] On the basis of results from Cape Crozier, young (three- to seven-year-old) males spend significantly more days incubating eggs than adult

Table 5.15 Length of shifts (mean number of days) for each member of a nesting pair during normal incubation routines.

		Shift				
	Year	First (male)	Second (female)	Third (male)	Number of Shifts	Reference
Ross Island						
Cape Royds	1959	11	10	6	8	44
	1961	8	2	?		43
	1962	14	12	?		54
	1964	9	8	?		54
	1965	13	11	?		54
Cape Bird	1969	15	11	?		40
Cape Crozier	1968	14	11	5	7	5
Cape Hallett	1959	14	16	5	3	31
	1960	13	14	4	<5	>4
Signy Island	1950	13	15	8	3	38
	1981	14	13	5	3	21
King George Island	1981	13	9	?	4	46
	1982	14	12	?	3	46
	1984	12	10	?	6	46
	1985	11	9	?	7	46
	1986	12	10	?	4	46

Table 5.16 Average number of days between first departure and next arrival among Adélie penguins of various ages, Cape Crozier, 1967–1974.

Age (yr)	Male Breeders (mean ± SE)	n	Male Nonbreeders (mean ± SE)	n	Female Breeders (mean ± SE)	n	Female Nonbreeders (mean ± SE)	n
2			16.9 ± 1.8	10			18.4 ± 2.1	17
3			24.9 ± 0.8	109	15.4 ± 1.7	16	24.8 ± 1.0	76
4	19.6 ± 2.5	14	32.3 ± 1.0	139	18.6 ± 1.1	36	29.3 ± 1.5	44
5	16.1 ± 0.8	77	31.4 ± 0.8	159	17.9 ± 0.7	102	30.1 ± 1.7	44
6	15.4 ± 0.3	207	27.3 ± 0.9	89	16.8 ± 0.3	237	27.5 ± 2.5	44
7	15.5 ± 0.3	177	28.0 ± 1.5	26	16.8 ± 0.2	218	29.0 ± 2.3	5
8	14.3 ± 0.5	61	24.0 ± 2.9	5	17.0 ± 0.3	98	28.3 ± 6.5	3
9	15.1 ± 0.8	25	21.0 ± 3.8	6	16.6 ± 0.8	31	27.5 ± 3.7	4
10	14.6 ± 0.5	31			16.6 ± 0.8	24		
11	14.9 ± 1.6	20	20.3 ± 0.5	3	16.6 ± 0.8	14		
12 and 13	14.1 ± 0.8	16			16.1 ± 0.4	24		
Total	15.3 ± 0.2	628	29.0 ± 0.4	546	16.8 ± 0.2	800	26.8 ± 0.8	237
r	−0.7612*		−0.9630*		−0.8598*		−0.6630	

$P < .05$.

(control) males ($P < .05$; t test), and young males spend significantly more days incubating eggs than male mates of young (three- to seven-year-old) females ($P < .05$). Consequently, young females spend significantly more days incubating eggs than female mates of young males ($P < .05$), but young females spend about the same number of days incubating eggs as control females ($P < .05$).

Thus young males differ from their older counterparts in that they are left on the nest longer during incubation. This could be so because the young males completed feeding trips to sea more rapidly than the average control male or mate of a young female, or their mates needed more time for feeding trips than the average control female (and, in fact, than the average young female). A comparison of the amount of time breeding birds of different ages spent at sea after their first incubation watch reveals that younger birds took more time to return, but there was little difference after five years of age (table 5.16). Thus because Adélie penguins tend to pair most often with birds of the same age (table 5.6), the second choice (longer feeding times by the young males' mates) seems to be the more accurate interpretation.

REVERSED INCUBATION ROUTINE

On occasion, the male Adélie penguin leaves first and the female takes the first incubation shift. In the majority of such cases, the male leaves before the second egg is laid. These are called reversed incubation routines. Rarely have reversed incubation routines been reported in the northern part of the species's range (Signy Island, King George Island).[21,38,46] Farther south they occur regularly but at low incidence: Point Géologie, 22 percent of nests in 1950;[36] Cape Royds, 17 percent in 1959, 2 percent in 1961;[44] Cape Bird, 8 percent in 1969;[40] Cape Crozier, 10 percent in 1968, 6 percent in 1969.[4] They are more prevalent when sea ice is extensive and compacted (hence their rarity in the northern part of the range) and are more common near the end of a colony's egg-laying period. The latter pattern follows from the fact that they occur often when the male has arrived early but paired late (i.e., has been fasting for a very long time before acquiring a partner who then lays eggs). In many of such cases, the female is young and the male old. Finally, reversed incubation schemes usually succeed when the male returns in less than about seven days. If he stays away longer, the female probably will desert the nest to seek food at sea, leaving the eggs to skuas.

At Cape Crozier, the frequency of reversed incubation routines varied little in relation to age (of one known-age partner), except in pairs that included a three-year-old female (table 5.17). The latter incubated first in six of sixteen instances. Otherwise, males usually (about 88 percent of instances) incubated first. The high proportion of three-year-old females taking first incubation duties probably followed from their late arrival date (ch. 4), which increased their chances of encountering unpaired males that had been present for a long time already. This was the case in almost all instances among known males who left first; that is, these males had arrived early and paired late. By the time they paired and courted, their fat reserves were low enough that they left prematurely, usually just as their mates laid the first eggs.

The Crozier data support Sladen's[38] contention that a normal (male first) incubation routine is necessary for successful breeding as a general rule.[5] In 73 percent of reversed incubation routines (sixteen of twenty-two), the female deserted before the male returned. In only 18 percent of

Table 5.17 Proportion of Adélie penguin males and females of known age taking the first incubation shift, Cape Crozier, 1963–1975.

Age (yr)	Male Percentage	n	Female Percentage	n
3			37.5*	16
4	94.1	17	17.1	35
5	87.8	82	15.0	100
6	89.1	313	12.3	284
7	87.4	286	15.1	265
8	84.7	157	14.0	143
9	86.5	52	22.5	40
10	87.5	48	25.0	28
11	97.3	37	8.7	23
12	93.1	29	10.5	19
13	81.8	11	14.3	7
14	100.0	2		
Total	88.1	1,034	14.9	960

*Statistically different from the combined percentage for all other age groups (14.5%; $P < .05$, t test).

the twenty-two cases did the pair raise at least one chick to fledging; in two other cases the males returned to relieve the female, eggs were eventually hatched, but no chicks were fledged. Upon taking the first incubation watch, females remained alone on their nests from two to fourteen days before deserting, the mean being 7.9 days for those deserting before their mate returned and 5.6 days for those relieved before desertion. Taylor[44] recorded a mean of 6.5 days (range: three to fourteen days) for four similar cases. The mean of 7.9 days appears to be a good estimate of the average physiological reserves available to a female after egg laying, although some endure as long as eighteen days. Of course, the difference depends partly on nutritional reserves upon arrival at the colony, energy expended in pairing and egg laying, and length of time at the colony before egg laying. Given that most experienced females pair within a day of arrival and that the average interval between arrival and laying is seven days (table 5.13), the average female who lays two eggs apparently can remain at the colony for about two weeks (7.0 plus 7.9 days). Only in years when the pack ice is dispersed can an appreciable portion of pairs who reverse their incubation routines successfully incubate their eggs to hatching.[4] In those years, the male is able to feed and return to relieve his mate in a short time.

■ Hatching Success

Hatching success is the proportion of eggs laid that eventually hatch. Determining hatching success and comparing it between studies may seem to be a simple task, but it is not. This is particularly true for a species that must incubate the eggs very closely to prevent freezing and maintain warmth. Thus the eggs remain well hidden from view by researchers. In light of this difficulty it is reasonable in any comparison to consider only the most divergent values, as in the discussion of clutch size. Besides the Adélie penguins' proclivity to conceal its eggs, various researchers have used different methods to assess egg loss and hatching success. This adds to the difficulty in making comparisons. Sladen[38] estimated egg and chick mortality merely by assuming that 1,791 nests began with two eggs and later counted the chicks in nests after hatching. Although this procedure produces results useable for some applications and reduces disturbance, other researchers have used more labor-intensive methods to assess the actual number of eggs laid.

Hatching success has been investigated at seven colonies spanning twenty-one nesting seasons for this species. Considering these studies, among all eggs laid, an average of 69 percent hatched (table 5.10). Only four times has hatching success been measured as significantly lower than this mean: at Ross Island, Cape Royds in 1962 (37 percent) and Cape Bird in 1968 (57 percent) and 1977 (56 percent); and at King George Island, Admiralty Bay, in 1982 (53 percent). At Cape Royds in 1962, the fast ice remained in place well into January, so breeders had a 45-kilometer trek to open water.[54] Mates took a long time to return to relieve their partners (table 5.15), so nests were deserted and eggs lost. Similarly, daunting ice conditions were implicated as a cause for greater desertions and the lower hatching success evident at Cape Bird in 1968;[40] reasons for the lower hatching success in 1977 are not known.[11] The lower success at Admiralty Bay in 1982 also was related to long incubation spells, but in this case long trips were thought to be related to a lack of food in nearby waters.[46] On the other hand, a lot of ice as at Ross Island has the same effect as reduced food supplies. Both situations increase the effort and time needed to find food and restore body condition.

Conversely, only six times has hatching success significantly exceeded the average: Cape Hallett in 1960 (82 percent), Cape Royds in 1964 (88 percent), Cape Bird in 1967 and 1970 (both 80 percent), and Signy Island in 1980 and 1981 (82 percent and 81 percent, respectively). There was very little sea ice in the vicinity of Cape Royds in 1964,[54] so trip times were short; the extent of sea ice at Cape Bird in 1967 and 1970 was much less than in 1969.[40] Presumably trips to and from the sea were short, leading to few desertions by mates. On the other hand, at Signy Island, the ice in 1981 was much more extensive than in 1980. Nevertheless, in this area, where sea ice cover is never as extensive as in the Ross Sea at the start of the Adélie penguin nesting season, hatching success was not affected. Conditions at Cape Hallett in 1960 were not reported.

CAUSES OF EGG LOSS

Determining the cause of egg loss or hatching failure takes great effort, vigilance, and some educated guesses. The event is rarely observed because researchers have not had the resources to watch even one nest continually for five weeks. Usually researchers use the circumstances of loss as clues to guess the cause. For example, if a bird remains alone at the nest for longer than the normal incubation shift or attends the eggs less closely, and then the nest is found deserted, it is assumed that the partner's failure to return was responsible for the loss. If the Adélie penguin in question never appears again, the egg probably is lost because of his or her death (or, as in one substantiated case, emigration). If he or she returns to the nest site later, it is assumed that tardiness caused desertion by the mate (who got hungry). In contrast, if a bird attends the nest alone for significantly less time than the average incubation watch and then the nest is found empty, that bird probably is responsible for the loss, although exact circumstances (e.g., skua predation, attack by another penguin, or simple desertion) are unknown. In some cases, eggshell fragments, a bloodied penguin, or other such clues add further evidence.

Some disparity between studies in the relative importance assigned to various causes of egg mortality results from variation in researchers' definitions of circumstances, particularly with respect to predation. The majority of authors believe that predation by skuas is rare and that most loss of eggs to skuas is the result of scavenging.[22,38,40,44,45,48,56] Scavenging occurs when nests are deserted or left unattended during fights, or eggs are kicked from nests. However, a few authors consider any egg loss to skuas predation.[11] This subject is discussed further in chapter 7.

Taylor[44] split mortality into segments by time or stage of breeding (i.e., eggs, guarded chicks, crèched chicks) and estimated 34 percent mortality of eggs (including 10 percent addled) during the 1959–60 season at Cape Royds. Yeates[54] estimated 32.5 percent and 43–52.5 percent overall egg-to-fledging mortality at Cape Royds for 1964–65 and 1965–66, respectively. Egg mortality was 12 percent during the first year and 37 percent the second. Causes of egg loss have been treated in greatest detail at Cape Bird.[40] Over a four-year period, 27.7 percent of all eggs laid failed to hatch: 6.7 percent were infertile, 7.6 percent were lost through nest desertion, 0.7 percent were poorly incubated, 0.4 percent were lost during fights, 2.6 percent were taken by skuas, and 9.7 percent could not be accounted for. Among the losses for which the cause was known, 79 percent occurred during the first twenty days of incubation. During a later season at Cape Bird (1977–78),[11] 43.8 percent of the eggs laid failed to hatch: 18.3 percent taken by skuas and 15.3 percent lost caused by nest desertion. At colonies in Lützow-Holm Bay, East Antarctica, loss rates were similar in 1990 to those observed at Cape Bird in the 1960s: desertion 20 percent, predation 4 percent, accidents (e.g., fights) 1 percent, and infertility 6 percent.[47]

All birds in these studies were of unknown age, although some authors[44] classified birds by experience, as deduced from nest site characteristics and time of laying. Egg mortality in this case at Cape Royds varied from about 35 percent in early-laying nests situated in the interior of colonies to about 45 percent in thirty-three interior and forty-three peripheral nests with late-laid eggs. A study at Cape Bird[11] found that loss caused by desertion did not differ between central (13.4 percent) and peripheral nests (16.6 percent), but loss to skua predation did vary by location: peripheral 27.1 percent, central 5.9 percent. Also, desertion was much more likely in one-egg clutches (most of which are laid late) than in two-egg clutches. As noted earlier, time of egg laying is affected strongly by the age of the parents, with young, inexperienced parents laying later and using more peripheral nests than older, more experienced birds.

The cause of egg loss is not constant over the incubation period.[11] Losses occur throughout the incubation period, but the loss to predation (or scavenging) by skuas is highest at the start (days 2–4), especially from

Table 5.18 Average number of eggs hatched as a function of the age and experience of Adélie penguin parents, Cape Crozier, 1963–1974.

	First-Time Breeders				Prior Breeders					
Age (yr)	Male (mean ± ± SE)	n	Female (mean ± SE)	n	Male (mean ± SE)	n	Female (mean ± SE)	n	Total* (mean ± SE)	n†
3			1.2 ± 0.1	29					1.2 ± 0.1	29
4	1.5 ± 0.1	11	1.4 ± 0.1	77			1.5 ± 0.1	8	1.4 ± 0.1	108
5	1.5 ± 0.1	36	1.5 ± 0.1	49	1.4 ± 0.2	7	1.5 ± 0.1	28	1.6 ± 0.1	252
6	1.5 ± 0.1	34	1.3 ± 0.1	21	1.8 ± 0.1	14	1.6 ± 0.1	29	1.6 ± 0.1	340
7	1.7 ± 0.1	16	1.6 ± 0.2	7	1.8 ± 0.1	12	1.6 ± 0.1	16	1.7 ± 0.1	285
8					1.4 ± 0.2	8	1.9 ± 0.1	11	1.6 ± 0.1	72
9							1.8 ± 0.2	6	1.8 ± 0.1	30
10–13					1.8 ± 0.1	30	1.8 ± 0.1	21	1.7 ± 0.1	81
Total‡	1.5 ± 0.1	97	1.4 ± 0.1	183	1.7 ± 0.1	71	1.6 ± 0.1	119	1.6 ± 0.1	1,197

*Regression of eggs hatched to parent age, $r = 0.8202$ ($P < .05$).
†Includes birds of unknown sex.
‡The mean for all first-time breeders, $1.43 ± 0.1$, was significantly different from that of all prior breeders, $1.64 ± 0.1$ ($t = 4.20$, $P < .05$).

nests with one-egg clutches. Loss caused by desertion usually does not occur until after the first week of incubation but peaks around day 22 and again near day 35. These patterns of desertion are consistent with early (day 8) losses being caused by reversed incubation routines that fail (female deserts within a week if the male's return is tardy) and losses peaking later in normal incubation routines caused by failure of the female to return from her first trip to sea (day 22) or of the male from his first trip to sea (day 35). Failure to return usually is the result of predation by leopard seals (see ch. 7).

As mentioned earlier, the degree of egg loss varies annually, which means that the causes vary as well. For instance, at Cape Crozier during a year of very extensive and concentrated sea ice (1968), a condition that slowed the journeys of breeders to and from the colony, 25 percent of breeders deserted their eggs. The next year, when winds were strong and the polynya well developed, making travel easier, only 10 percent of nests were deserted.[4] However, the age of the birds was important: different age groups lost eggs for different reasons. Therefore this subject is developed further in chapters 6 and 7.

EFFECT OF AGE AND EXPERIENCE

At Cape Crozier, egg mortality decreased with increased age of breeders. The average number of eggs hatched was lowest among pairs containing a four- or five-year-old male (1.4–1.5 eggs) or a three- or four-year-old female (1.2 to 1.5 eggs; table 5.18). For pairs in which the known-age male or female was older than five years, an average of 1.7 eggs hatched per nest. These figures, when compared with those on number of eggs laid (table 5.11), indicate egg mortality of 11 percent in pairs where penguins were five years of age and older, about 14 percent where males were four years old, and about 20 percent where females were three or four years old. Breeding experience also had an effect on the number of eggs hatched or lost (table 5.18). On average, first-time breeders lost 0.3 to 0.4 eggs per nest, but birds with at least one year of previous experience lost only 0.1 to 0.2 eggs per nest ($P < .05$, t test). No significant difference emerged between age and experience groups of birds in the chronology of egg loss.[5]

ADDLED EGGS

A proportion of eggs are not lost but nevertheless fail to hatch. Although incubated to full term, 10 and 7 percent, respectively, of eggs failed to hatch in studies at capes Bird and Royds.[40,44] Failure to hatch can result from any of several causes. Fertilization may never occur because of lack of or improperly timed copulation, infertile spermatozoa or ova, or immotile or otherwise disabled spermatozoa. Fertilized eggs may cease to develop because of freezing, fluctuating temperatures of incubation, or inherent abnormalities in the embryo. Any of these factors could result from behavioral or physiological insufficiencies related to the age (maturity) of parent birds. For example, incomplete copulations (no sperm transferred) have been observed in both gentoo[8] and Adélie penguins.[2,38]

In the yellow-eyed penguin, eggs are larger and more likely to be fertile as females get older.[32,33] In addition, fertility decreases in very old females. In the study of Adélie penguins at Cape Crozier,[5] parental age significantly affected the frequency of infertile eggs (see table 5.18; $r = -0.9608$, $P < .05$). The eggs of three- and four-year-olds showed the highest percentages of infertility (25 percent), compared with 22 percent in five-year-olds and 13 percent in six- and seven-year-olds. The latter proportion was similar to that of adult controls. The figures for young Adélie penguins are comparable to those for three-year-old yellow-eyed penguins,[33] in which 30 percent of eggs are infertile. Incredibly, 68 percent of eggs laid by two-year-old yellow-eyed penguins failed to hatch because of infertility. In Adélie penguins, because of the dangers inherent in breeding, especially among young birds, such a low fertility would quickly select against breeding at so young an age.

■ Further Thoughts

In this chapter, we have considered ecological aspects of nesting through the egg stage, but that is only half the story. The discussion of chicks and breeding success in chapter 6 will pave the way for further discussion of the ecological significance of breeding patterns.

Chapter 6

The Reoccupation Period

Chicks and Breeding Success

Adults who lose their eggs return to reoccupy their nest sites at about the time that eggs successfully incubated by other pairs begin to hatch, hence the name *reoccupation period* for this phase of the breeding cycle.[49] Breeders who failed early in the breeding process will remain for a few weeks. Coincidentally, young nonbreeders begin to visit in a sequence structured by age: younger ones arrive the latest and stay the shortest amount of time. These birds are actually accomplishing their first occupation rather than reoccupying the colony. Many are returning for the first time since they fledged two to four years earlier. Their arrivals swell the colony population, and the colony again takes on an air of busy, noisy penguindom. In contrast, the adults that are still tending chicks are more sedate. The main activity of breeders revolves around guarding the chicks and making repeated trips to sea to find food for them. Once the chicks are large enough and strong enough to resist skuas on their own, parents spend little time in the colony and almost no time at the nest site. In effect, these parents' short visits reduce the apparent number of adults using the colony site at this time. These parents will continue to return until their chicks disappear, either by gaining independence (fledging) or being taken by skuas.

In this chapter, we consider the fate of breeders and their charges.

■ The Chick

NEWLY HATCHED CHICKS

Once the egg is pipped, the Adélie penguin chick emerges from the shell within twenty-four to sixty-eight hours; in one sample of fifty-two eggs, nine chicks took less than twenty-four hours to hatch, thirty-eight took twenty-four to forty-eight hours, and five chicks took more than twenty-four hours.[43,49,55] A pipped egg is one in which the chick has used its beak to drill a small hole. Like the chicks of most other avian species, the Adélie penguin chick has a small, hard, and sharp bump near the end of its beak. It is white and is called the egg tooth. It disappears within a few weeks after hatching.

Newly hatched chicks, as measured at Cape Hallett in 1960, averaged 86.3 grams;[44] one at Cape Adare weighed 85.7 grams.[31] The newly hatched chick can barely hold its head up and spends its first three to four days tucked completely under the parent. It does not need to feed right away because it is nourished by a large yolk sac within its body cavity. In

Illustration 6.1 An Adélie penguin parent with a chick just a few days old. *Drawing by Lucia deLeiris © 1988.*

the egg, the yolk sac is external to the embryo, but by the time of hatching it is covered by the chick's body wall. The sac at hatching averages 13.1 grams (range 3.9–17.2 grams) and ranges from 6.4 to 24.2 percent of body mass.[44] It is completely absorbed by the time the chick is about a week old.[42,43] The majority of Adélie penguin chicks (70 percent) are fed during the first day after hatching, but some (that eventually survive) are not fed until the fifth day. Without being fed, depending on the size of its yolk sac, a chick can survive on average 6.4 days, although after about five days it probably is incapable of regaining its condition.[44] The yolk sac enables the chick to await the uncertain date of arrival of a parent fresh from the sea with food;[49] delays usually are caused by severe ice conditions. Therefore the yolk sac is an important adaptation of the Adélie penguin.

CHICK PLUMAGE

The newly hatched chick is completely covered with down, ranging in color from pale silver to dark, sooty gray.[49,55] This first down gives way to a woolly second down, so that by about ten days of age all chicks are the same dark gray color. Thereafter, chick plumage develops as follows:[55]

20–25 days: Loss of second down begins underside of flippers.

25–30 days: Underside of flippers is mostly down-free; down disappears from legs and lower abdomen.

30–35 days: Tail quills show; legs, lower back, lower abdomen, and chin are down-free.

35–40 days: Back is mostly down-free; patches of down are missing from face, breast, and chin.

40–45 days: Down is gone from under chin, above beak, around eyes, and most of back and breast.

45–50 days: Some chicks are completely down-free; others have small tufts top of head, neck, and back of flippers.

50–55 days: Odd down tufts remain, mostly on top of the head; chicks depart for the sea.

CHICK GROWTH

The mass of chicks increases with age from an average 95 grams at hatching[24] to a maximum at about forty-two to forty-five days of age[7] (fig. 6.1). Chicks then lose about 15 percent of mass before fledging at about

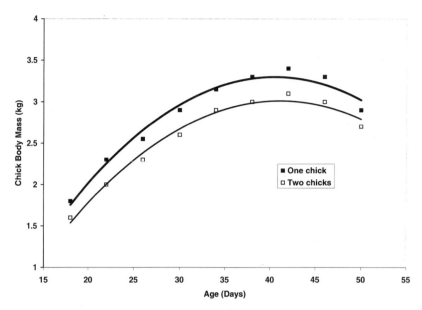

Figure 6.1 Growth curves of Adélie penguin chicks from one- and two-chick broods at Cape Crozier, 1969–1970. *Redrawn from Ainley and Schlatter.*[7]

fifty-two days of age. The rate of growth and the maximum mass attained vary by locality, season, brood size, and age of parents. At Cape Crozier in 1969–70, average maximum mass of single chicks reached 3.3 kilograms, and chicks from two-chick broods reached 3.1 kilograms.[7] These values averaged 0.7 kilograms less than maximum mass of chicks at Cape Royds in 1959–60.[56,57] This difference is not surprising. Given the large amount of variation in ice conditions and food availability, similar and even larger differences in maximum chick mass can occur from year to year at any one site.

Fledging mass can also vary. Over a seven-year period at Admiralty Bay, average fledging mass regardless of brood size ranged from 3.2 kilograms in 1981 and 1986 to 3.8 kilograms in 1977.[28,60] The mass of fledglings, from single-chick and two-chick broods, respectively, averaged 2.9 and 2.7 kilograms at Cape Crozier in 1969–70 and 3.4 and 3.2 kilograms at Cape Royds in 1959–60.[7,57] Maximum mass and fledging mass at Gorlay Peninsula, Signy Island, in 1950–51 were similar to the values for Cape Royds: 4.3 and 3.7 kilograms, respectively.[49] At fledging, the Cape Crozier chicks were 75 percent and the Cape Royds chicks 89 percent of the mass of adult females weighed in mid-January.[7]

The rate of growth and ultimate fledging weight of chicks differ depending on both the size of the brood and the age of the parents.[7] The youngest parents, three years of age, produce chicks that on average are 71 percent of the body mass of chicks fledged by parents who are six years and older (table 6.1). The older parents also raised two chicks to fledging more often than young parents. Whether young parents bring less food to their chicks or bring it less frequently than older parents is not known. As noted in chapter 5, younger parents took longer to return from foraging trips at sea during the incubation period. Therefore, they may be feeding their chicks less frequently, because their foraging trips last longer than those of older parents.

■ Phases of Parental Care

GUARD STAGE

During the first few weeks of life, the chick never leaves the nest and is always attended by a parent. This phase in parental care is called the guard stage.[45,49] The parents change roles every one to three days. Males are present with the chicks on 52–55 percent of days, females 45–48 percent.[5,55] The age of a parent does not appear to affect its share of guarding duties.[5]

Table 6.1 Average body mass of chicks caught on the beach just before departing for sea and the size of broods fledged by Adélie penguin parents of known-age, Cape Crozier, 1969–70.

	Chick Mass		Two Chicks Fledged	
Parent Age (yr)	Kilograms	n	%	n
3	2.0	1	0	4
4	2.4	9	8	25
5	2.6	17	20	41
6	2.9	18	18	54
7	2.8	14	24	29
8	2.8	13	36	14
Controls	3.0	22	28	43
r^*	0.8720		0.9627	

*Regression related to age of parent ($P < .05$).

Chicks are strong enough to stand by themselves at six to ten days, and begin to move about the territory at eleven to fifteen days.[51] Until this age, if the chick does leave the nest (perhaps kicked out during a fight), there is little likelihood that it will ever return because it then becomes an intruder to be repulsed by all territory holders, even its own parents. Between eight and seventeen days of age, parents and chicks learn to recognize one another, mostly by voice.[59] Parents will not allow a chick back into the nest unless they can recognize the chick's voice. It is not until about twenty-two days of age that chicks leave the nest on their own.[51] By this time, parents must be able to recognize their chicks, and chicks must know where their nest is located in case they need to return (i.e., for feeding).

The guard stage lasts, on average, twenty-two days (range sixteen to thirty-four days).[21,55] However, length of the guard stage depends on when a chick hatched relative to others in the subcolony, with early-

Illustration 6.2 An Adélie penguin parent standing beside its crèche-age chick. *Drawing by Lucia deLeiris © 1988.*

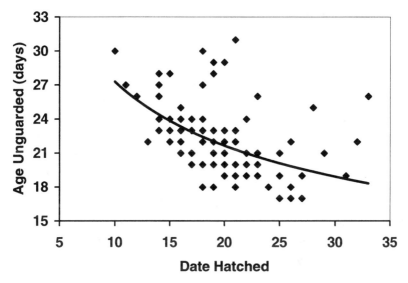

Figure 6.2 The age of chicks when they enter the crèche as a function of the date hatched. *Data from Taylor.*[56]

hatching chicks being guarded longer than late-hatching ones (fig. 6.2). The result of this timing is that the colony's breeding schedule becomes even more synchronous than during egg laying. Even though hatching is spread over about four weeks, the end of the guard stage (beginning of crèche stage) is spread over only two weeks.[55] The increased synchrony may protect crèched chicks from skua predation.

CRÈCHE STAGE

When left alone, chicks seek the company of other chicks and form groups know as crèches. The period when chicks are independent of the nest is known as the crèche stage.[45,49] A crèche is defined as three or more chicks closer to one another than half the internest distance.[21] Mean age at which chicks enter the crèche (same as termination of the guard stage) was determined to be twenty-two days at capes Bird and Royds (range sixteen to thirty-four days)[21,55] during single years of study and 23.1 ± 0.2 days at Cape Crozier over four years.[5] Age at crèching ranged sixteen to nineteen days at Signy Island, depending on whether the chick was single (nineteen), or was the first- (seventeen) or second-hatched (sixteen) member of a brood.[32]

Murray Levick, who thought erroneously that parents feed any chick in their breeding group, gives an interesting perspective on the process and purpose of crèche formation:

> Whilst the chicks are small the two parents manage to keep them fed without much difficulty; but as one of them has always to remain at the nest to keep the chicks warm, guard them from skuas and hooligan cocks, and prevent them from straying, only one is free to go for food. Later on, however, two other factors introduce themselves. The first of these is that the chick's downy coats become thick enough to protect them from cold without the warmth of the parent; and the second that as the chicks grow they require an ever-increasing quantity of food, and at the age of about a fortnight this demand becomes too great for one bird to cope with. At this time it is still necessary to prevent the chicks from straying and to protect them from the skuas and "hooligans," and so to meet these two demands a most interesting social system is developed. The individual care of the chicks by their parents is abandoned, and in place of this, [sub]colonies start to "pool" their offspring, which are herded together into clumps or "crèches," each of which is guarded by a few old birds, the rest being free to go and forage. (p. 96)[31]

We know now (building on Levick's observations) that parents leave their chicks unattended when the chicks reach a stage of development at which they can thermoregulate (at fifteen days),[25,26,46] the chicks' growth rate can be sustained only if both parents are foraging simultaneously, the chicks and parents can recognize one another (at eight to seventeen days),[59] and chicks can find their way back to the nest site.[51]

Leaving the chicks may not be an entirely altruistic strategy on the part of parents, but may be a defense against persistent begging by the chicks. James Murray, naturalist on Shackleton's first expedition, made some interesting observations that indirectly illustrate some of these points:

> A bird was taken from a nest which had a chick in it and put down at a little distance. Meantime the chick was put in a neighbour's nest. Presently the bird came running up. It started back on seeing the empty nest, not in alarm or fear, but exactly as if thinking, "I've come to the wrong house!" and trotted off to a distant part of the [colony]. Her reasoning seemed to be this: "There was a chick in my nest, therefore this

empty nest cannot be mine." She couldn't imagine the chick leaving the nest, and so never searched for it. It was only a yard from the nest all the time. After half an hour's searching in vain for any place like home she returned to the nest, and accepted the restored chick as a matter of course. (cited in Shackleton p. 355)[48]

Several other factors also affect the age at which a chick enters the crèche. First, during years when food becomes difficult to find, the relationship shown in figure 6.2 becomes much steeper, meaning that late-hatching chicks enter the crèche at younger ages than normal and may be left prematurely. The result is greater risk of predation by skuas. Second, parents nesting in locations vulnerable to skuas tend to remain longer with their chicks. Thus the relationship is less steep.[55] Vulnerable nests are those that are isolated or located at the edge of a subcolony. Finally, the parents' age affects how long they guard the chick. The chicks of older parents (ages nine to thirteen years) enter the crèche when twenty-one to twenty-three days of age, compared to twenty-three to twenty-six days of age for chicks of youngest parents (three to five years).[5] The pattern of older age at crèching among chicks of young parents is contrary to the one in which later-hatching chicks are left alone at younger ages because younger parents lay their eggs later than average. Together, these patterns indicate a role on the part of the chick in determining when it enters the crèche, suggesting that it is not merely the act of parents leaving that determines the beginning of crèching. Perhaps the chicks of young parents develop more slowly as a result of being fed less food and thus take longer to acquire the urge to wander from the nest.

The number of chicks raised to the crèche increases with a parent's age up to eight years of age, as shown in the study at Cape Crozier (table 6.2). The chicks of older parents also weigh more than those of younger parents, as noted earlier, thus giving them greater chances of survival once they fledge. Finally, previous breeding experience is a factor: parents that have bred before raise more chicks to crèche than those observed breeding for the first time (table 6.2). In most penguin studies, the number of chicks raised to crèche versus raised to fledging is used as the measure of breeding success because it is so difficult to follow (i.e., locate day after day) the chicks once they enter the crèche. However, studies have shown, that once a chick has reached the crèche, mortality becomes very low.[23,55]

The first chicks left unattended in a subcolony have difficulty in finding refuge because they are considered intruders by parents still guarding

Table 6.2 Average number of Adélie penguin chicks raised to crèche by parents' age and breeding experience, Cape Crozier, 1967–1974.

	First-Time Breeders				Previous Breeders					
Age (yr)	Male (mean ± SE)	n	Female (mean ± SE)	n	Male (mean ± SE)	n	Female (mean ± SE)	n	Total* (mean ± SE)	n
3			1.3 ± 0.1	13					1.3 ± 0.1	13
4	1.6 + 0.2	7	1.3 ± 0.1	42			2.0	3	1.4 ± 0.1	52
5	1.6 ± 0.1	24	1.5 ± 0.1	37	1.5 ± 0.1	2	1.4 ± 0.1	17	1.5 ± 0.1	80
6	1.6 ± 0.1	23	1.4 ± 0.1	17	1.8 ± 0.1	12	1.6 ± 0.1	21	1.6 ± 0.1	73
7	1.8 ± 0.1	13	2.0	3	1.7 ± 0.1	10	1.7 ± 0.1	12	1.7 ± 0.1	38
8–13					1.8 ± 0.1	33	1.7 ± 0.1	33	1.8 ± 0.1	66
Total†	1.6 ± 0.1	67	1.4 ± 0.1	112	1.7 ± 0.1	57	1.6 ± 0.1	86	1.6 ± 0.1	322

Calculations based on chicks that hatched, not eggs laid.
*Regression of age versus number of chicks to crèche, $r = 0.9568$ ($P < .05$).
†First-time breeders (1.48 ± 0.1 chicks) are less successful than breeders with previous experience (1.64 ± 0.1 chicks; $t = 2.853$, $P < .05$).

their own chicks. The parents attempt to fend off the visiting chicks. These chicks, especially if from nests near the edge of a subcolony, become more vulnerable to skua predation (ch. 7). They find refuge by successfully running underneath the pecks of a parent whose chicks are so large that the parent cannot effectively peck over or around them. The unguarded chick then becomes temporarily adopted, in effect, although the foster parent will not feed the new arrival. Soon, however, enough chicks are moving around that territorial defense against chicks by parents becomes fruitless and they no longer bother. The number of chicks in an Adélie penguin crèche averages 12.5 ± 0.9 (range three to twenty).[21,32]

Chicks form crèches for protection from the elements as well as protection from skuas.[21,49] As long as the chicks remain well within the boundaries of the subcolony, the presence of failed or nonbreeders, who are maintaining territories, is enough to deter most skuas. Crèche formation usually coincides with the peak in the numbers of adults (mostly nonbreeders) present during the reoccupation period. Some early naturalists described these adults as guardians of the crèche because they rushed out after any skua that came too close to the subcolony. What was really happening was that the nonbreeders were repulsing skuas from the vicinity of their own nests, regardless of whether they had eggs or chicks of their own to defend. In a crèche, besides the protection afforded by the mass of many chicks grouping closely together and inadvertent protection by nonbreeders, the chicks also enjoy safety in numbers. As Murray noted,

> When the young birds are well grown, if there is an alarm they flock together, and any old birds present in the [sub]colony form a wall of defence between the young and the enemy [skua]. This habit has given rise to the belief that they are somewhat communistic in their social order, and that defence of the [sub]colony is a concerted action. It is not so. Each bird is defending its own young one only, and will often fight with another of the defending birds, or peck at any young one which comes in its way. (cited in Shackleton, p. 357)[48]

Besides protection against predators, crèches are also important as protection against the elements. The tightness of a crèche is directly related to temperature. When it is warm and chicks are panting, there are no groupings of chicks; when it is blustery, the chicks push together, forming densely packed clumps.[21,73]

■ Chick Survival

Fledging success is measured as the proportion of successfully hatched chicks that survive to fledge. Average fledging success for the Adélie penguin is 77 ± 2 percent, tallied from twenty-one seasons of study at seven localities (see table 5.10). In general, fledging success is significantly higher than hatching success, perhaps reflecting the greater investment in reproduction on the part of parents once the egg stage is successfully negotiated. Estimates of fledging success were significantly lower than average in six instances. At Admiralty Bay, King George Island 1982 (68 percent), food supply was thought to be unusually low.[60] At Cape Royds 1962 (72 percent),[73] Cape Crozier 1968 (70 percent),[5] Signy Island 1980 (47 percent),[32] and Lützow-Holm Bay 1990 (70 percent),[65] ice conditions led to unusually long foraging trips by breeding birds (see table 5.15). The conditions that brought lower success at Cape Bird in 1977 (63 percent) are not known.[23] Conversely, fledging success significantly surpassed the average in six instances: Cape Royds 1965 (90 percent),[73]

Cape Bird 1967 (83 percent),[50] Cape Crozier 1968 (96 percent),[5] Admiralty Bay 1977 (83 percent) and 1984 (84 percent),[60] and Signy Island 1990 (97 percent).[32] In these cases, parental foraging trips were short, at least compared with other years in the various studies (see table 5.15). Presumably food supplies were easy to exploit. In the cases of cape Crozier, Bird, and Royds, as well as Admiralty Bay, the instances of high fledging success followed particularly low hatching success. Thus some sort of compensatory process may have been involved, with parents taking better care of fewer chicks.

The causes of chick mortality are even more difficult to ascertain than the causes of egg mortality. There is much disparity in authors' rankings because of biases in interpretation. There is agreement that skua predation of chicks is more of a problem than it is for eggs, mainly because parents have more difficulty protecting chicks than eggs (ch. 7). Indeed, the loss of chicks to skuas is much higher in peripheral than in central nests because the skuas must attack at ground level.[23,50,55] The difficulty comes in interpreting what is predation and what is, in effect, euthanizing a weak, starving chick (alive but not really capable of protecting itself, with no likelihood of fledging). There is also disagreement about the causes of starvation. Chicks that are deserted eventually starve, lose body mass and strength, and quickly become susceptible to skua predation. The fact that more second (smaller) chicks are lost than are first or single chicks[23] indicates that starvation can be an issue (because second chicks do grow less than the others). Some authors[22,23] contend that starvation rarely, if ever, results from inadequacy of the food supply available to parents. Rather, chicks don't get fed because of deficiencies in the behavior of the parents (e.g., parent loss, inadequate coordination of foraging trips). On the other hand, chicks are smaller, more susceptible to predation, when less food is delivered to them.[8] When food is more difficult to procure, parents also leave their chicks less well protected, which also increases susceptibility to skuas. Finally, mass starvation of chicks caused by a failed food supply has been noted at certain colonies. For example, this was the case at Béchervaise Island in 1995 and 1998,[27,29] but this is a rare occurrence.

The results of one study at Cape Hallett that examined the cause of death of chicks that did not disappear may help explain some of the broader categories in table 6.3.[44] In that study, 28 percent of dead chicks were squashed; at least some of these chicks probably were squashed by fighting adults. Of the remaining dead chicks, 11 percent were deserted

Table 6.3 Causes of chick loss by Adélie penguins compared between research projects, by percentage.

Location	P	D	S	E	F	O	Reference
Cape Royds, 1959	68	19		8		5	55
Cape Bird, 1967–70	45	30			5	20	50
Cape Bird, 1977	63		35			2	23

P = predation; D = desertion; S = starvation; E = exposure; F = fighting by adults; O = other/unknown.

(and starved), 39 percent starved without being deserted, and for 22 percent the cause of death could not be determined.

The rate of chick mortality is not constant as chicks grow and mature. Once chicks reach twenty-five to thirty days of age (i.e., after several days in the crèche), mortality becomes negligible until fledging. Chicks weigh about 1.9 kilograms (single-chick broods) or 1.6 kilograms (two-chick broods) when they enter the crèche[7] (note that these values are from Cape Crozier in a year of low chick mass compared with other studies, as mentioned earlier). According to one study at Cape Bird, skuas cannot overcome a chick that is 3.0 kilograms or heavier, and 3.0 kilograms is attained by about thirty days of age.[21,74] Predation reaches a maximum around ten days of age,[23,50] which is about when chicks become too large to be brooded tightly. Predation continues to be a significant hazard through twenty-five days. Starvation, at least that caused by loss of a parent or poor coordination between parents, is most important near the end of the first week of life. Before then, the chick lives on the reserves contained in its yolk sac. In rare circumstance, when food is difficult for adults to procure, few parents remain long in the colony. Then, not only do chicks become lighter and weaker, but few adults remain to fend off skuas. In those cases, skuas can take chicks almost at will, including chicks much older and more vigorous than usual.

■ Fledging

The Adélie penguin chick leaves the colony on its own, unattended by parents, when it is sixty (first chick of a brood) or sixty-one (second chick) days old, on average, according to one study;[32] fifty-two days old according to another;[63] and fifty-one days (range forty-one to fifty-six) according to a third.[55] Age at fledgling is younger for late-hatching chicks than for early-hatching chicks.[55] The latter pattern is consistent with that of fledging age being a function of parental age. Recalling that younger parents lay later, the chicks of young parents (three to four years of age) fledged at 47.2 ± 1.5 days of age ($n = 29$), compared with 49.1 ± 0.4 days of age for chicks of older parents (five to eight years of age; $n = 140$ nests of which one parent was of known age). The difference is on the verge of being statistically significant ($t = 1.797$).[5]

Chicks sometimes fledge within a few days of being fed by a parent, so there is no indication that they must be starved before they will want to leave. In fact, Taylor[56] observed that well-fed chicks are almost indifferent to parents' visits just before fledging. Indeed, the better fed they are, the better chance they have of surviving to find food, which they do on their own. Fledglings leave the beach in small groups and sometimes swim back to shore after the first plunge. They are particularly vulnerable to leopard seals during their first swims, being unaware of the danger (ch. 7).

Illustration 6.3 Adélie penguin chicks almost ready to fledge. *Drawing by Lucia deLeiris © 1988.*

Levick's notions about fledging were erroneous but colorful. Had his observations been made at a colony large enough to attract leopard seals (see ch. 7), no doubt his conclusions would have been different. I believe he was observing insatiable, begging chicks pursuing their parents onto the beach and into the water:

> In the autumn of 1912, at a small rookery which I came upon on Inexpressible Island, I had an opportunity of watching their first attempts in this direction. Crowds of young Adélies were to be seen on the pebbly beach below their rookery, much of the ice having disappeared at this late season, leaving bare patches of shingle which were very suitable for the first swimming lesson.
>
> Many old birds paddled in for a short distance, and crouching in a few inches of water, splashed about with their flippers to give the youngsters a lead. Some of the latter needed little encouragement, and took readily to the strange element, very soon swimming about in deep water, but others seemed more timid, and these latter were urged in every possible way by the old birds, some of whom could be seen walking in and out of the water, and so doing what they could to give their charges confidence.
>
> In this duty one or two old birds might be seen with a little crowd of youngsters, so that evidently the social instincts which gave rise to the crèche system in the first place were extended to the tuition of the young and thus to their preparation for the journey north.
>
> Up in the rookery, fully fledged youngsters could be seen clamouring in vain for food, the old birds resolutely refusing to feed them now that they were able to forage for themselves. (p. 112)[31]

■ Breeding Success

Clutch size, hatching success, and fledging success combine to make up breeding success. It is measured as the percentage of eggs that result in a fledged chick or the number of chicks fledged per breeding pair of adults. On average in this species, 51 ± 2 percent of eggs produce a fledged chick, as measured during twenty-six seasons at seven sites; or 1.0 ± 0.1 chicks are fledged per breeding pair, as measured in thirty-four seasons at nine sites (see table 5.10). The percentage of eggs that produced a chick significantly exceeded the average during nine seasons and sites: Cape Royds 1964 (68 percent) and 1965 (57 percent),[73] Cape Bird 1967

(66 percent) and 1970 (64 percent),[50] Cape Crozier 1968 (65 percent),[5] Cape Hallett 1960 (67 percent), Admiralty Bay 1984 (59 percent) and 1986 (61 percent),[60] and Signy Island 1985 (60 percent).[18] Except for Signy Island in a year for which there were no supporting data, these cases were years in which rates of both hatching success and fledging success exceeded respective averages. The reasons for elevated hatching and fledging success in these years, as discussed earlier, usually were related to favorable sea ice. In only five cases has breeding success been recorded that is significantly lower than the average: Cape Royds 1962 (26 percent),[53] Cape Bird 1968 (43 percent) and 1977 (36 percent),[23,50] Admiralty Bay 1982 (36 percent),[60] and Signy Island 1980 (38 percent).[32] All these rates are very low compared with the average, and in all cases except Signy Island, poor hatching success was primarily responsible for the reduced breeding success. Eggs hatched poorly in years of extensive sea ice, which means that eggs were lost as nests were deserted by birds waiting too long for their mates to return from the sea.

A more conservative estimate of overall breeding success is chicks fledged per pair (c/p) of the pairs that attempted breeding (laid eggs). Values of this measurement exceeded the average in only five cases: Cape Crozier 1968 (1.3 c/p);[5] Signy Island 1981 (1.3 c/p);[32] and Magnetic Island 1982 (1.6 c/p), 1986 (1.4 c/p), and 1987 (1.5 c/p).[67] The values for Magnetic Island seem to be unusually high compared with those calculated at other sites, and it is possible that the method used to calculate them differed greatly from those used elsewhere. In any case, I have no explanation for the elevated rates. In the case of Cape Crozier and Signy Island, elevated rates occurred during seasons of sparse sea ice, which as noted earlier is beneficial to this species. In contrast, values were significantly lower than the average in seven cases: Cape Bird 1968 (0.7 c/p)[50] and 1977 (0.7 c/p),[23] Admiralty Bay 1982 (0.7 c/p),[60] and Béchervaise Island 1991 (0.7 c/p), 1994 (0.1 c/p), 1995 (0.4 c/p), and 1998 (0.4 c/p).[27] In most cases, sea ice was much more extensive than usual, especially during the incubation period. This was especially so for Béchervaise Island, which normally is locked in by several kilometers of fast ice through most of the nesting season. Thus a slight reduction in prey abundance can exacerbate an already difficult situation. The mass of meals fed to chicks in the poor seasons at Béchervaise Island was much lower than in more productive seasons. Moreover, foraging trips by parents were longer than usual, which means that chicks were fed less food less frequently than the average. The result was that chicks starved.

CENTRAL VERSUS PERIPHERAL NESTERS

Several authors have noted differences in the productivity of seabirds nesting at the periphery of colonies compared with those nesting more centrally (i.e., with at least one nest between theirs and the periphery[50,51]): black-headed gulls (*Larus ridibundus*),[39] gannets,[35–37] kittiwakes,[13,72] and royal penguins.[10] Most of these authors concluded that centrally nesting individuals were better breeders and therefore better able to compete successfully with other individuals for central territories. Only Brian Nelson[36] disagreed. He concluded that young birds predominate among peripherally nesting gannets because they arrive later than older birds and that the lower reproductive success of peripheral birds resulted from age-related factors rather than qualitative differences between central and peripheral birds. Nelson came to this conclusion because he could recognize age classes of breeders by virtue of plumage differences, and he noticed that older birds tended to nest at their site of previous years rather than moving to more central positions left vacant because of

Table 6.4 Reproductive success of known-age Adélie penguins as a function of nest location in subcolonies at Cape Crozier.

	Peripheral Nest			Central Nest		
Age (yr)	Successful (%)	Chicks/Nest (mean ± SE)	n	Successful (%)	Chicks/Nest (mean ± SE)	n
1967, 1968						
3 and 4	41.3*	0.5 ± 0.1†	46	63.4*	0.8 ± 0.1†	41
5	50.0	0.7 ± 0.1	54	52.2	0.8 ± 0.1	46
6 and 7	52.4	0.7 ± 0.1	42	48.5	0.7 ± 0.1	33
1974						
5	60.4	0.8 ± 0.1	53	65.2	1.0 ± 0.1	46
6 and 7	65.2	0.1 ± 0.1	141	74.1	1.0 ± 0.1	139
8	70.6	1.0 ± 0.1	34	72.7	0.9 ± 0.1	33
9–13	69.2	1.0 ± 0.1	65	75.7	1.1 ± 0.8	70
Total	60.0*	0.8 ± 0.1†	435	67.6*	0.9 ± 0.1†	408

*†Difference statistically significant ($P < .05$, t test).

the former owner's death. Such vacancies were filled by young gannets who returned early enough to find these sites unoccupied.

A number of researchers who did not have banded Adélie penguins of known age and breeding experience with which to work have spent much effort demonstrating that breeding success is higher in pairs that laid eggs early and were in the centers rather than at the periphery of breeding groups. The pattern was proved to be true especially in regard to hatching and fledging success.[23,40,50,58] Such early, central nesters are older birds, so the results are not surprising. In fact, unlike young, first-time breeders, older birds nesting at the periphery of subcolonies apparently can compensate for their position, perhaps by being more attentive to their eggs and chicks (table 6.4).

The fact that older Adélie penguins avoided both peripheral nest sites and sites more than three nests toward the center of the subcolony is particularly important in regard to the methods of study used by researchers who compared nesting success between "central" and "peripheral" nesters. Richard Tenaza[58] explored the problem in greatest detail and stated that "most central nests evaluated were actually near colony edges (to minimize observer disturbance)" (p. 81). His motives were exemplary but resulted in a comparison between one sample with the highest possible proportion of young birds (i.e., peripheral birds) and another sample with the highest possible proportion of oldest birds (i.e., birds nesting one or two sites away from the periphery). Little wonder that he discovered major differences in nesting behavior and success between his two samples. Others[23,40,50] did not define "central" nest in their respective samples. Because these researchers were sensitive to disturbance, however, it is likely that they, too, studied samples similar to that of Tenaza.

Thus it appears that the Adélie penguin is similar to some other seabirds in regard to the relative significance of peripheral and central nest sites. The North Atlantic gannet is another species that is conservative in its tendency to switch from one nest to a more central one, and large numbers of old birds do nest at subcolony peripheries.[36] There are also some parallels with the kittiwake.[72] However, these species do not seem to avoid central-most nests in large subcolonies. Because these birds fly to their territory, they do not have to trespass on other territories to reach their own, as Adélie penguins do. The fact that Adélie penguins avoid nests in the center of large subcolonies may explain why young birds avoid certain subcolonies.[5] About 50 percent of subcolonies at the large Cape Crozier colony were composed of fewer than fifty nests, and in such subcolonies few nests would be more than four from the periphery.

EFFECT OF AGE ON BREEDING SUCCESS

The ineptitude of young birds is reinforced when they pair together. In combinations involving three- and four-year-olds at Cape Crozier, the tendency to produce only one egg was high. Out of seven such pairings, only two (29 percent) resulted in two eggs being laid, and in both cases the three- or four-year-old paired with a bird five years or older. In fifty-five pairings involving older birds, fifty-two (95 percent) produced two-egg clutches.[5] In raising chicks to crèche age, any parental age combination including a breeder four years old or younger performed below the population average, but age combinations in which both breeders were five years or older performed at or above the population average (cf. tables 6.5, 8.5). These patterns may be consistent with that in the gannet, where young birds that would not otherwise have bred did so on the rare occasion that they paired with a much older partner.[37] Those younger gannets could also be high-quality individuals, as in the case of kittiwakes.[72]

DURATION OF PAIR BONDS AND EFFECT ON BREEDING SUCCESS

The number of seasons that a pair remains together is known to influence reproductive success in several seabird species; birds retaining the same mate average higher success than those that change.[12,33,45,71] For the

Table 6.5 Mean number of chicks raised to the crèche for Adélie penguin pairs in which both partners were of known age, Cape Crozier, 1967–1974.

Age of Bird	Age of Partner				
	4	5	6	7	8+
3		0.0 (1)	0.0 (1)		
4	0.5 ± 0.5 (2)	0.5 ± 0.5 (2)			
5		1.5 ± 0.5 (4)	1.3 ± 0.3 (3)	1.0 ± 0.6 (3)	
6			0.7 ± 0.3 (6)	0.8 ± 0.6 (3)	1.7 ± 0.3 (7)
7+				1.1 ± 0.3 (7)	2.0 ± 0.0 (4)

Sample size in parentheses.

Table 6.6 Average clutch size of Adélie penguins with respect to number of seasons paired with the same partner and prior breeding experience, Cape Crozier 1964–1975.

	Number of Seasons Bred		
Seasons with Same Mate	1	2	3+
First	1.5 ± 0.1 (101)*	1.7 ± 0.1 (43)	1.8 ± 0.1 (36)
Second		1.8 ± 0.1 (39)	1.8 ± 0.1 (18)
Third+			1.9 ± 0.1 (10)

Sample size in parentheses.
*Statistically smaller than all other categories ($P < .05$, t-test).

Adélie penguin, changing mates at the Windmill Islands was found to have no effect on overall breeding success (percentage of pairs that fledged chicks) in one season (1959–60) but had a definite negative effect in another (1960–61).[40] In the second season much of the effect resulted from the failure of split or disunited birds to pair in time to breed. Over four seasons at Cape Bird, changing mates had no effect on overall productivity among birds that bred after the mate change but, again, a high proportion of birds who did not retain their previous mates failed to breed.[50,52] All researchers who have studied the phenomenon of faithfulness to mate in Adélie penguins agree that asynchrony of return and death of a partner are the major factors leading to split and disunited pair bonds, respectively (see ch. 4).

At Cape Crozier, clutch size increased markedly only in the second year of breeding with the same mate and only for second-time breeders (table 6.6). Retaining the same mate beyond the second year had a slight but statistically insignificant effect. In terms of overall breeding success, the average number of chicks fledged per breeding pair did not increase, but the percentage of pairs fledging at least one chick did (table 6.7). Thus it appears that Adélie penguins find little advantage in retaining the same mate from one season to the next. Because asynchrony of return and death are the main reasons Adélie penguins lose their mates, being able to switch to a mate with similar synchrony would be advantageous. In so doing, Adélie penguins continue to mate with birds of the same age throughout their reproductive period. Keeping the same mate apparently is advantageous only because it eliminates the need to compete again for

Table 6.7 Number of chicks fledged and percentage of Adélie penguin pairs fledging at least one chick with respect to mate retention and number of seasons bred, Cape Crozier, 1963–1974.

	Number of Seasons Bred			
Seasons with Same Mate	1	2	3+	Total
First	1.0 ± 0.1 (101)	0.9 ± 0.1 (36)	0.9 ± 0.2 (24)	1.0 ± 0.1 (161)
	75.2%	58.3%	58.3%	68.9
Second+		1.0 ± 0.2	1.2 ± 0.2 (18)	1.1 ± 0.1 (56)
		68.4%	83.3%*	73.2%

Sample size in parentheses.
*Percentage breeding successfully statistically significant ($P < .05$).

a new one.[3,40,50,52] However, mortality among breeding birds is so high that the likelihood of losing a mate is also high.[4]

Mate retention and its consequences therefore follow a different pattern in Adélie penguins of the southern Ross Sea than in other species of long-lived seabirds or even in Adélie penguin farther north (at the Windmill Islands, Wilkes Land).[40]

■ What Breeding Patterns of Related Species Tell Us About Adélie Penguins

We can learn much about the ecological problems that confront an Adélie penguin and how it addresses them by comparing them with other species. This is especially true of closely related species such as the gentoo and chinstrap penguins.

The Adélie penguin's range overlaps the ranges of its two closest relatives, the gentoo and chinstrap penguins, in a narrow zone located in the northern part (mostly the northwestern coast and outlying islands) of the Antarctic Peninsula (60° to 64°S). At a few locations within this zone, such as Signy Island and King George Island, these species nest side by side. Comparison of the natural history patterns among these species at these special locations provides insights into the ecology of the Adélie penguin that otherwise would not be apparent. Here we review the findings of comparative studies as a way to summarize and elucidate the ecology of the Adélie penguin (table 6.8).

Table 6.8 Comparison of aspects of the breeding ecology and natural history patterns of pygoscelid penguins.

	Adélie	Chinstrap	Gentoo
Nesting habitat	Hummocks	Slopes, high	Flats, low
Internest distance (cm)	43	60	74
Average body mass, M–F (kg)	4.50–4.20	4.30–3.70	5.50–5.06
Egg volume (cm^3)	99–106	92–93	?
Egg mass (g)	111–121	112–114	124–130
Mass egg 1:egg 2	Heavier	Equal	Lighter
Egg:female body mass (%)	2.64	2.65	2.58
Incubation period (days)	33.5–35.6	33.4–36.4	35.3–37.6
Date of peak hatching	Dec 4	Dec 28	Dec 13
Chick mass at hatching (g)	95	75	85
Days between hatching	1.5	0	
Feeding frequency, guard (per day)	1	1.4	2
Feeding frequency, crèche (per day)	2	2.9	3.8
Average food loads of parents (g)	420–650	400–500	680–920
Chick age at crèche (days)	16–19	23–29	20–33
Crèche size (no. chicks)	10–20	2–3	?
Fledging period (days)	50–55	55–60	62–105
Growth constant, *K*: Gompertz	0.066	0.065	0.057
Growth constant, *K*: logistic	0.146	0.127	0.113
Fledgling, percentage of adult mass	75–89	89	104
Chick age at fledging (days)	51	52	72
Place of molt	Pack ice	Colony	Colony
Migratory	Yes	Partly	No

Data from several studies.[7,24,32,43,57,60–64,68–70]
As much as possible, estimates were made within the zone of breeding overlap.

There is a consensus among researchers that the differences in the breeding biology of Adélie, chinstrap, and gentoo penguins largely reflect each species' adaptations to conditions in the center or larger part of its respective range.[32,61] Adélie penguins nest in the harsh conditions of high latitude, where they must cope with sea ice and a short summer. The gentoo nests mostly on subantarctic islands, which are mild with a long summer season; sea ice and snow-covered ground are not an issue. The chinstrap exists in the maritime portion of the northern part of the Antarctic zone. Sea ice is present only a short time during the summer; otherwise, conditions are not severe.

Natural history patterns differ depending on whether the penguin species is migratory.[5,17] Sedentary species tend also to feed inshore, make short foraging trips, breed at an early age, and have high degrees of breeding site and mate fidelity. Migratory species tend to feed offshore, make long foraging trips, first breed at older ages, have higher mate separation rates, and produce smaller clutches.

The nonmigratory gentoo penguin feeds close to the colony, mates change nesting duties every few days, and individuals never fast during the nesting season. They nest in flat areas near the sea and molt where they nest. The prelaying activities of mated pairs can last for several weeks before eggs are laid. Peak laying is about two weeks later than that of Adélies nesting at the same location. The gentoo's eggs are smaller relative to adult body size, but incubation takes longer (thirty-eight days first egg, thirty-five days second), and chicks hatch at a lower body mass (85 grams). In general, smaller eggs should take less time to hatch, but apparently the gentoo does not incubate its eggs as closely as does the Adélie penguin.[43] Gentoo chicks grow slowly (growth constant, K, is 0.113 or 0.057, depending on what curve is used to describe growth over the life of the chick), remain in the guard stage longer, and fledge at an older age (mean seventy-two days, range sixty-two to eighty-two days). At fledging, gentoo penguins chicks are equal in body mass to adults.

The chinstrap penguin is migratory to the extent that it is absent from

Illustration 6.4 Gentoo penguin. *Drawing by Lucia deLeiris © 1988.*

Illustration 6.5 Chinstrap adult and chick. *Drawing by Lucia deLeiris © 1988.*

nesting colonies when sea ice is present, which is only for a few months at the latitudes where it nests. It spends its nonbreeding period at sea, north of the sea ice. It lays eggs a month later than the Adélie penguin, a timing that coincides with disappearance of sea ice in the spring. When sea ice remains unusually late in the spring, egg loss in the chinstrap penguin is very high because this species cannot survive the long fasts typical of the Adélie penguin.[32] Also in years of extensive, long-lasting sea ice in spring, the numbers of chinstraps that attempt breeding is substantially lower, in a pattern opposite to that of the Adélie penguin (see chs. 2 and 9).[61] Chinstraps lay the smallest eggs relative to body size among the pygoscelid species, and the chick hatches at the lightest mass (75 grams). The first and second eggs are similar in size. After laying the eggs, females weigh the same as males and always take the first incubation turn. Incubation shifts of three to eight days, depending on year (and ice conditions), are much shorter than those of the Adélie. Both eggs of a two-egg clutch hatch at the same time. Chicks are guarded more than a week longer than are those of the Adélie and fledge a week older. Growth rate is similar to that of the Adélie chick, which is higher than the rate for gentoo chicks. Fledglings are 89 percent of adult mass, compared with 79 percent for Adélies and 100 percent for gentoos. Chinstraps molt at the colony, including yearlings who arrive at the colony to molt, mostly at the end of the breeding season.

The breeding patterns of the Adélie penguin are very different from those of the gentoo and, to a lesser extent, those of the chinstrap penguin. Most of the differences are keyed toward allowing the least amount of time necessary to occupy colony sites and maintaining contact with sea ice. This is most evident at either end of the nesting cycle. Adélie penguin courtship is very short, shorter even than the amount of time it takes to form eggs; egg laying is a month earlier than it is for the chinstrap and two weeks earlier than for the gentoo; and Adélies molt mainly in the pack ice away from the colonies. Adélie penguin egg size is the largest, newly hatched chicks weigh the most, the guard stage and the entire chick stage are the shortest, and chicks fledge at the youngest age, weighing the least relative to adult size. Obviously, the Adélie penguin's reproductive strategy depends on its chicks being more likely to find food than those of the other species. The latter may be ensured by finding sea ice.

Some important aspects of the Adélie penguin's breeding patterns clearly are adaptations to extensive sea ice early in the nesting season: storing and physiologically mobilizing fat to survive long fasts and laying a clutch in which the eggs are markedly different in size and hatch a few days apart, thus facilitating brood reduction, if necessary. In such cases, the younger, smaller chick loses out to the larger, stronger sibling in competition for their parents' attention. The larger crèches of Adélie penguins probably are responses to the colder conditions they encounter and perhaps the fact that chicks form crèches at an earlier age and thus are more vulnerable to skuas than the other pygoscelid species. Finally, the nesting habitat of the Adélie (high on hummocks) is an adaptation to avoid drifting snow and meltwater; the other two pygoscelid species nest in gentler terrain (see also ch. 3).

The severity of conditions in which the Adélie penguin nests thus dictates much of its breeding activity. Some conditions are beyond adaptation, but they illustrate the miraculous toughness of this creature. Levick offered these telling observations of conditions that neither a chinstrap nor a gentoo would ever encounter:

> During both spring and summer there are occasional snowstorms, and during these the birds sit tight on their nests, sometimes being covered up by drift. As a rule the bird on the nest keeps a space open by poking its head upwards through the snow, but sometimes it becomes completely buried. Air diffuses so rapidly through snow that death does not take place by suffocation, and the bird can live for weeks beneath a drift, sitting on its nest in the little chamber which it has thawed out by its own warmth. Generally after a few hours the snow abates and settles down sufficiently to expose the nest once more, but sometimes a breeze springs up which is not strong enough to blow the snow away, but simply hardens the surface of the drift into a crust which last for several weeks, and the birds are imprisoned in consequence. Then little black dots are seen about the surface of the drift, which are the heads of penguins thrust through their breathing holes.
>
> On one such occasion I witnessed an interesting little incident. An imprisoned hen was poking her neck up through her breathing hole when her mate spied her and came up. He appeared to be very angry with her for remaining so long on the nest, being unable to grasp the reason, and after swearing at her for some time he started to peck at her head, she retaliating as far as her cramped position would allow. When she withdrew her head, he thrust his down the hole till she drove it out again, and as this state of things seemed to be going on indefinitely, I came up and loosened the crust of snow which imprisoned her, on which she burst out, and seemed glad to do so. She was covered with mire, having for many days been sitting in a pool of thaw water which had swamped her nest and evidently spoilt the eggs. When I put her back on the nest, she sat there for some time, but eventually they both deserted. (pp. 105–106)[31]

In fact, the problem of drifting snow is becoming more important to the Adélie penguin in the Antarctic Peninsula region, where climate amelioration is leading to greater snow fall (ch. 9).

■ Reproductive Effort: Age Versus Experience

In this and the preceding two chapters, it was amply demonstrated that behavioral and reproductive characteristics vary with age in the

Adélie penguin. This is also true in other seabird species in which age-related reproductive success and activities have been studied: yellow-eyed penguin,[45] little blue penguin,[20] albatross of several species,[66] northern fulmar (*Fulmarus glacialis*),[38] short-tailed shearwater,[47] Brandt's cormorant (*Phalacrocorax penicillatus*),[9] South Polar skua,[6] western gull (*Larus occidentalis*),[41,54] red-billed gull,[33] black-legged kittiwake,[16] and arctic tern (*Sterna paradisaea*).[15] In the studies of Adélie penguins, only a few aspects of reproductive biology bore no relationship to age, including incubation period and the interval between laying and losing eggs. Not all aspects of the Adélie penguin's reproductive ecology have been investigated for age-related variation, however. For example, nothing is known about the size of eggs laid by known-age female Adélie penguins. Egg size is known to vary with age in other seabirds, such as yellow-eyed penguin, black-legged kittiwake, and red-billed gull.[11,34,45] It would be surprising if this were not true for the Adélie penguin.

The following factors have been shown to vary with age in the Adélie penguin: arrival date in the spring, ability to overcome unfavorable sea conditions during spring migration, body mass and amount of subdermal fat present upon arrival,[1] amount of time spent in the colony during the breeding season, number of visits to the colony, activities at the colony (i.e., wandering, occupying territory, pairing, and breeding),[5] frequency of displaying (for certain displays),[2,3] facility in social interactions,[3] breeding incidence, amount of time between spring arrival and egg laying, date of egg laying, nest quality, nest location within the subcolony, clutch size, incubation routine, duration of feeding trips to sea, egg loss, prevalence of infertile eggs, number of eggs hatched (as a function of egg loss and infertility), nestling survival, age of the chick when left alone by parents, fledging success, body mass of fledglings, site and mate fidelity,[30] and survivorship.

It also has been found that in many of these factors, sex and experience at the colony or in breeding modified the relationship to age. For example, males arrived earlier in spring and spent more years as prebreeders and nonbreeders after breeding once, and females bred at an earlier age and suffered higher mortality. One and sometimes two (but usually not more) years of previous breeding experience increased breeding success through effects on many of the factors. Therefore experience at the colony or in breeding was the primary factor affecting many age-related changes in this species's breeding ecology; some of the described changes probably resulted from added experience at sea.

Using other seabird species as a guide (e.g., yellow-eyed penguin,[45] little blue penguin,[20] and black-legged kittiwake[14,72]) it is surprising that the accumulation of breeding experience had little effect on the behavior and reproductive success of Adélie penguins beyond one year. The frequent establishment of new pair bonds in Adélie penguins of all ages—often the factor responsible for the lowered reproductive success of newly established pairs in other seabird species—may have masked the effects that superior experience could have in oldest Adélie penguins. Few data have been compiled to test the effect of very long pair bonds in the Adélie penguin, but extended pair bonds certainly have been found to increase breeding success in other species.[12,14,19,45,72] Also contributing to the lack of an extended experiential effect in Adélie penguins must be the extreme hazards of breeding, which are unrivaled by any other seabird that has been adequately studied over the long term. Adélie penguins do not have time to accumulate experience slowly.

Chapter 7

Predation

Most of the smaller seabird species, such as storm petrels, diving petrels, auklets, and murrelets, often are preyed on by certain larger birds, such as frigatebirds, gulls, skuas, eagles, and falcons. Unlike the adults of most large-bodied seabirds, therefore, these smaller seabirds are not at the apex of the food web. In response to such predation, these smaller seabirds have acquired activity patterns that bring them to breeding colonies only during the night. In the very highest latitudes, where it is daylight twenty-four hours per day during summer, few of these smaller species nest. At the locations where most nest, there is at least twilight, if not one to two hours of darkness, which is when they become active at the colony. With the dawn, evidence of the predation is apparent around the nests, in the pellets cast by the predators, or in the carcasses found at their roosts. How the predation came about is mostly obscured by the darkness. The victims may have bumped into the wrong animal during the night. Only falcons, such as peregrines (*Falco peregrinus*), take these seabirds as prey during the day, and they do so also on the wing at sea. On occasion, fish take these smaller, diving species as well.

Therefore it is anomalous that penguins, among the largest (and certainly the heaviest) of all seabirds, also find themselves beneath the apex of the marine food web pyramid. Indeed, much lore surrounds the predator-prey relationships between pinnipeds and penguins, especially the leopard seal and the Adélie penguin. All happens in broad daylight within a stone's throw of observers, beginning with naturalists in the heroic days of Antarctic exploration, who have described the seal-penguin interaction. The relationship between skuas and penguins, especially the South Polar skua and Adélie penguin, has also inspired much interest and discussion. However, the latter sort of predator-prey relationship is not unique to penguins. Rather, most small- to medium-sized seabird species must deal with gulls or skuas, which lurk about nesting colonies looking for anything edible, including eggs and small chicks.

In this chapter I review the relationships between Adélie penguins and their supposed or actual predators: the killer whale (*Orcinus orca*), leopard seal, and skua. This information should provide a context in which to better understand the Adélie penguin, especially its demography and the spatial and temporal variations in breeding success.

■ Killer Whale

In Antarctic lore, the killer whale is considered to be a predator of Adélie penguins. The source of this idea is obscure, but Walker's text and other summary texts of mammals list seabirds among the prey of this large predator: "When they see a bird, seal, or other food, near the edge of the ice, they will dive deeply and rush to the surface breaking ice that is 1 meter thick and dislodging their prey into the water" (p. 1122).[35] Although many observations have confirmed such predation on Antarctic seals,[18] I have yet to find one with respect to penguins. An entry in the log of the *Terra Nova*—the ship used in Scott's second expedition—by its captain, Commander Harry Pennell, provides some indication about how the penguin-killer whale lore came about:

> Adélie penguins do not seem to mind the killer whales unless very close, for though at times they will come shooting out of the water on to a floe in evident hurry and a killer appears soon afterwards, yet at other times they will remain in the water quite close to a whale. It seems as if the killers tried to get them under the ice, and not attack them in open water. (cited in Lowe and Kinnear, p. 111)[16]

Thus in this instance there was no observation of the killer whale catching or eating an Adélie penguin, but some connection between the two was inferred. The fact that the penguins were observed "shooting out of the water" is no proof because they almost never leave the water in any other way.

Killer whales do prey on the largest penguin species, but this does not seem likely for the smaller species, except on rare occasions. An Adélie penguin could swim circles around a killer whale in most circumstances, and one would not provide much nourishment. The whale would expend much more energy in the capture than it could secure by eating the bird. Reports indicate that killer whales on very rare occasions do kill and sometimes consume smaller penguins such as African, macaroni, and rockhopper, typically in the vicinity of seal rookeries (where the whales are actively hunting seals).[5,36] On the other hand, the recurring take by killer whales of fledglings of the king penguin is well documented as the latter first swim from the breeding island.[9,10] In this case, however, the penguin prey is five to six times the size of an Adélie penguin, the victims are clustered together, they have high caloric value (being very fat), and in taking their first swims they are exceedingly slow and naïve. Therefore it makes sense that killer whales would show more than passing interest. These penguins are definitely worth the effort in caloric terms. On many islands king penguins were brought to near extinction, along with elephant seals (*Mirounga leonina*) and whales, by people acquiring animal oils during the eighteenth and nineteenth centuries.[7,29] The single penguin found among hundreds of killer whale stomachs inspected from the Southern Ocean[18] was an emperor penguin.[26]

Killer whales are known to "play" with seabirds. They have been observed capturing cormorants and then drowning them while mouthing and nudging them.[36] Similarly, killer whales are known to play with Adélie penguins. In one case observed at Cape Crozier in 2000 (G. Ballard, personal communication), a pod of four or five killer whale calves practiced herding single Adélie penguins. They chased the penguin several hundred meters, and at times different individuals spurted ahead to cut off the penguin's changes in direction, much as a dog would herd sheep. This involved very vigorous activity on the part of the whales and the penguins. In the end, the penguins escaped by jumping onto ice floes. The whales herded four penguins in succession. In the mean time, large numbers of adult killer whales appeared to be diving beneath nearby fast ice, seemingly to feed (perhaps on fish), and thousands of other penguins swam to and fro in flocks in areas where the adult killer whales were present.

Adult killer whales were reported to chase Adélie penguins once before, but the details and outcome are unknown.[32] Otherwise, many observers have reported the close proximity of Adélie penguins to killer whales and have concluded that neither species cared in the least about the other. Typical are the observations of Levick:

> One day, as I watched some hundreds of Adélies bathing in an open lead, suddenly the back of an enormous killer-whale . . . rose above the surface as it crossed the lead from side to side, appearing from beneath the ice on one side and disappearing beneath it on the other. To my surprise, not the slightest fear was shown by the birds in the water. Had this beast been a sea-leopard, there would have been a stampede, and every bird have leapt from the water on to the sea-ice. On this evidence I formed the opinion that in all probability killer-whales do no harm to Adélie penguins; later I saw it confirmed, when a school of killers shaved close past several floes that were crowded with Adélies, and made not the least attempt to get at them, as they might so easily have done by upsetting the floes. Very probably this is because the agile bird can escape with such ease from the ponderous whale, and fears it no more than a terrier fears a cow, though he thinks twice before coming within reach of its jaws. (p. 88)[15]

I can only concur with Levick, having witnessed on numerous occasions the close proximity of many thousands of Adélie penguins and many dozens of killer whales, with no response by either toward the

Illustration 7.1 Killer whales hunting a crabeater seal. *Drawing by Lucia deLeiris © 1988.*

other. It would be a learned response, as African penguins actively avoid being in the water with these whales.[27]

■ Leopard Seal

The relationship between Adélie penguins and the killer whale is exactly opposite to that of the penguin and the leopard seal. As Levick writes,

> Some idea of the depredations committed by these animals [leopard seals] may be gathered from the fact that in the stomach of one which we shot I found the bodies of eighteen penguins, in various stages of digestion, the beast's intestines being literally stuffed with the feathers remaining from the disintegration of many more. . . . Evidence goes to show that the sea-leopard is the only living enemy, excepting man, that threatens the life of the adult Adélie penguin. (pp. 87–88)[15]

A summary of the scientific literature found nine studies that reported the remains of penguins in the stomachs of leopard seals.[22] Three of these studies specifically mentioned Adélie penguins. In a sample of 121 seals collected between 1901 and 1964, mostly near penguin colonies, 26 percent contained penguin remains. Among another sample of eighty-four leopard seals collected in the pack ice of the southern Scotia Sea and northern Weddell Sea, penguins were found in 16 percent, krill in 50–72 percent, fish in 12 percent, and squid in 9 percent. On the basis of these data and an estimate of the number of leopard seals in the Southern Ocean, Thor Oritsland[22] estimated that in a given year leopard seals consume 0.38 million metric tons of penguins per year, which translates to many thousands of penguins. Most of these penguins would be Adélies and emperors because these penguins and the leopard seal inhabit the sea-ice zone that rings Antarctica. Other penguin species live mostly in ice-free seas or at the periphery of the sea-ice zone (ch. 2), where leopard seals are rare.

The hunting behavior of leopard seals and predator avoidance behavior of Adélie penguins have been the subject of several studies, the most exhaustive of which were conducted at Cape Crozier, Ross Island[13,14,19,23] and at Magnetic Island, Prydz Bay.[28] These researchers stationed themselves on hilltops above the beach to observe most of its length. They could then see the coming and going of penguins and watch for seal pre-

Illustration 7.2 Leopard seal on an ice floe; note the large head and jaws. *Drawing by Lucia deLeiris © 1988.*

dation. Observations at Magnetic Island were supplemented by censuses of seals and penguins at the fast ice edge, some distance from the colony, using helicopters. More than 100,000 pairs of Adélie penguins nest at Cape Crozier and in the vicinity of Magnetic Island (Vestfold Hills, table 3.1), and in the season when the penguins are present, four to six leopard seals are also present at any one time. At a smaller colony, such as Cape Bird North (25,000 pair), two to four seals are present, and at tiny Cape Royds (5,000 pair), single leopard seals are seen, but only on occasion (personal observation[13]). Apparently, there must be enough penguins coming and going to keep a seal interested. Leopard seals also are rare at the tiny colonies in Terra Nova Bay, Victoria Land.

The number of seals at a given beach is limited because they space themselves along the beach a couple hundred meters apart. However, the seals may take turns moving into waters near the colonies from offshore areas of sea ice.[14] At Magnetic Island, nineteen seals were marked, and that summer, fourteen were resighted. Only four of these were present hunting penguins at any given time. Individual seals patrolled the beach at the penguin colony for periods averaging 1.9 days, then they disappeared for periods averaging 5.3 days.

SEAL PREDATORY BEHAVIOR

Unlike the Weddell seals also present, leopard seals have an aversion to coming ashore, even to rest. If leopard seals did come ashore, a territorial penguin might be easy prey if it stood its ground. I have seen elephant seals among Adélie penguin nests, and the penguins do stay put (except to avoid being squashed). Luckily for the penguins, leopard seals haul out only on ice floes or fast ice to rest or sleep. If penguins are present on a floe when a seal of any kind hauls out, they scatter quickly and maintain a space around the seal. A common sight is a seal sleeping on an ice floe surrounded by a halo of penguins, none no closer than about 5 meters. Even attempts by seals to grab penguins by lunging from the water at those waiting at floe edges or on the beach are futile. The penguins are just too quick and wary.

Leopard seals catch Adélie penguins mostly by being stealthy. Most often, the seals wait at floe edges or at the ice foot that forms along beaches at colonies. An ice foot begins to form in autumn (or winter at low latitudes) after the sea begins to freeze. Winds push sheets of sea ice against and onto the beach. The ice buckles, and the floes overlap. They freeze together and into the sand and gravel, becoming anchored. The result is a shelf of ice along the shore. The shelf is augmented by additional layers of ice floes each time the wind blows onshore during the winter; it is smoothed by the addition of drifting snow. Once the ice foot is formed, any floating sea ice abuts this shelf or foot and rises and falls with the tide. Having been heavily augmented with ice, this shelf remains long after most of the floating sea ice has disappeared. Leopard seals wait quietly at the ice foot with just their nostrils above the water surface. Lying in shallow water, they might submerge for two minutes or so, then emerge to breathe. This behavior is called patrolling. At low tide, a penguin might have to leap 2–3 meters to make it onto the ice foot. They can often do so, but if they fail and fall back into the water, many are taken by leopard seals.

An extreme example of the leopard seal's stealth was observed at Cape Bird, Ross Island[6] in 1995 (G. Miller, personal communication). There, during late summer, a large, muddy river flows across the colony and into the sea from a glacier nearby. In the runoff plume offshore, visibility is near zero. Leopard seals have been observed stationing themselves in the clear water at the edge of the plume, usually hiding alongside floating ice cakes. When a small flock of penguins arrives from sea, those that enter the plume become disoriented and begin to jump clear of the water, presumably to see where they are going. It is then that the seal dashes into the plume, seemingly locating the penguins by the large amount of splashing they produce in their maneuvering. Several successful captures were observed.

Adélie penguins are most vulnerable when the shore lead (open water between beach ice foot and offshore floating ice) has recently frozen and is covered by new ice, little more than consolidated slush. The penguins then venture across this ice, just thick enough to support their weight. The seals lurk below and hurdle upward, crashing through the soft ice to snare a penguin. The penguins are especially vulnerable when the ice foot is breaking up. At that time many ice chunks are washing to and fro in the surge of ocean swells (the appearance of which has led to the disintegration of the ice foot). Penguins hesitate to barge through because getting caught between ice blocks is very dangerous. In their hesitancy, looking for an ice-free way to shore, the penguins become easier prey to seals, which are lurking just offshore of the brash and cakes. Along a 200-meter stretch of the beach at Cape Crozier on January 2, 1966, William Emison found thirty-two dead or injured Adélie penguins: five had ruptured stomachs, fourteen had broken legs, and thirteen had seal-inflicted wounds.[23]

Leopard seals can capture Adélie penguins in open water, but the majority of penguins escape by out-swimming the seal. Capture takes a huge amount of effort on the part of the seal, eventually wearing the penguin out. The seal lunges at the bird as it twists this way and that, eventually making the capture with a final twisting thrust. A seal rarely has the energy to take two penguins successively in this manner. Using stealthy means, single seals have been observed making as many as fourteen captures in two days (nine in 110 minutes, five in 134 minutes).[28] During one seventy-minute period at Cape Crozier, a single seal captured seven penguins, of which it ate six.[13] Under good hunting conditions, the seals can catch enough penguins that eventually they begin to play with the victims. In such circumstances, penguins often escape with very severe wounds (most of which heal). This probably was the fate of the seventh penguin.

When a seal captures a penguin, it grasps it by the head and then flings it to break its neck. It then grasps its skin and flings the penguin about to remove the skin (and feathers). The skin strips off in pieces, which skuas fight over. Once it gets the skin off, the seal eats the very large breast or pectoral muscles. It is a pretty inefficient affair because only the seal's incisor and front teeth are designed for the task (the back teeth are more suited for straining krill). The bites can almost be described as dainty because the seal grasping the corpse again and again appears to be bobbing for apples. According to one study, it takes the seal seventeen to sixty-five bites over six to fifteen minutes to finish its meal.[19]

If a penguin does escape the seal and can still swim away, usually it has sustained horrific wounds. However, Adélie penguins are incredibly tough, and many survive. I have seen penguins with cuts through the skin into the muscle beneath, completely encircling their neck. These birds are colored red and black. But they just sit still for days, moving only their eyelids. Eventually, they are as good as new except for the scar. Fairly often, one sees scarred penguins or penguins with a portion of a wing removed.

Estimates of the number of penguins taken by leopard seals during the nesting season have been made only at colonies large enough to keep the seals interested. The estimates range from 1.4 to 5 percent of the penguin breeding population at Cape Crozier[19,23] to 2.7 percent at Magnetic Island.[28] As mentioned earlier, only occasionally do leopard seals hunt penguins at small colonies, such as Cape Royds or Inexpressible Island. Predation by leopard seals is also rare in Arthur Harbor, Anvers Island (14,000 pairs).[12] There the seals eat a lot of krill.

The number of adults taken in the pack ice away from colonies has not been counted. However, it must be significant given the low survival rate apparent in Adélie penguins. Although a significant mortality occurs at the colony, it is not high enough to account entirely for the depressed adult survival rate. Along ice edges (and perhaps among the floes of pack ice) leopard seals are far denser where Adélie penguins are concentrated.[28]

The leopard seals also prey on fledglings when the latter first venture into the water. The fledglings exhibit no predator avoidance and are easy prey to the seals. On the other hand, chick fledging is so synchronous that surprisingly few are taken. They enter the water in large numbers, but a seal can deal with only one at a time. One study at Cape Crozier, where perhaps 100,000 chicks fledge in a typical year, researchers estimated that 0.63 percent of fledglings were taken by the seals.[19] How long it takes the young penguins to learn that leopard seals are dangerous is not known. Until then, they are vulnerable. Leopard seal predation may be significant during some winters, when sea ice is unusually extensive.[38]

PENGUIN AVOIDANCE

Adélie penguins avoid being captured by leopard seals by adhering to certain rules.[19]

- Travel only in flocks, which enables penguins to detect seals more easily. The value of flocks or schools as a means of predator avoid-

ance was touched on in chapter 2 in the discussion of penguins as predators.
- Spend as little time as possible in the water, especially within about 200 meters of the beach. In this zone, Adélie penguins are always moving. Usually their movement is perpendicular to the beach. Only if they are certain that no seals are present will they swim slowly along the ice foot, but even then they remain 30–50 meters off. They do not pause to bathe after departing the beach until they are 200 meters or more from the beach.[1]
- Where there are ice floes, walk on them, jumping from one to the next.
- Avoid thin ice.

Illustration 7.3 A leopard seal contemplating some potential prey, which are safe for the time being. *Drawing by Lucia deLeiris © 1988.*

- When moving across thin ice, become completely motionless if a seal appears nearby or below. Eventually, the seal will go away. In one instance, observers noted a penguin on thin ice who did not move for at least eighty minutes.
- When a seal is encountered in the water, the group should scatter rapidly in all directions. This is the flash escape behavior mentioned in chapter 2.
- When at the water's edge, stay a few steps back until another flock of penguins swims safely by or comes ashore. Only upon observing their safety is it safe to dive into the water.

■ South Polar Skua

The relationship between the South Polar skua and the Adélie penguin is more complex than that between the leopard seal and penguin. This is because the skua acts as both a scavenger and a predator. The current consensus among researchers is that the skua is mostly a scavenger, taking eggs and chicks that would be lost anyway.[20,40,41] Among earlier naturalists, the skua was considered more of a true predator that took eggs or chicks at will.[17] However, it was recognized that most of the chicks taken were smaller and weaker than average.[30] One thought was that the skua actually helped the penguin, from an evolutionary standpoint, by ridding the gene pool of weaker individuals.[8,30]

Brown skuas also take eggs and chicks from Adélie penguins.[11,24,34] However, the interaction is limited to the very northern edge of the Adélie penguins' range because the brown skua inhabits mainly the Subantarctic, so their ranges do not overlap much. The larger brown skuas are dominant over the South Polar skuas and exclude the latter from the penguin colonies. The South Polar skuas therefore must eat fish in that situation. It is surprising that the larger skuas do not do more damage to the penguin colonies than they do. The brown skuas must be just as wary as their smaller cousins of the pecks and flipper beatings of an Adélie penguin.

Earlier naturalists thought that the South Polar skua depended on the Adélie penguin for food to raise its own chicks and that the number of penguins thus limited the size of skua populations.[8,15,17,30] Now it is known

Illustration 7.4 A South Polar skua picks at a penguin chick it has killed; the penguin adults are oblivious because the skua is out of their range. *Drawing by Lucia deLeiris* © *1988.*

that this is not true because many skua colonies exist independent of penguin colonies[2] and much of the skua's diet is composed of fish.[25,37,41] Furthermore, skuas must continue to raise their young for a month after the last Adélie penguin has left the colony for the sea. They can scavenge carcasses left behind, but that resource is limited, in some locations more than in others.[33] By nesting near penguins, skuas share a mutually needed resource—snow- and ice-free land—and by being where penguin eggs and chicks are available they acquire insurance against difficulty in procuring food at sea.[20] Even so, the first thing a skua eats from a penguin chick it has killed is the contents of the chick's stomach. Skuas also consume large amounts of fish and krill that are spilled as penguin chicks are being fed and that the skuas catch at sea.[3]

FACTORS THAT ENCOURAGE SKUAS

We are fortunate to have very detailed information on the relationship between the South Polar skua and Adélie penguin, thanks to the studies conducted and subsequent book written by Euan Young.[41] The book was based on twenty years of research. Young discussed the aspects of penguin behavior and biology that skuas take advantage of to obtain penguin eggs and chicks:

- Certain penguin behaviors distract the parents' attention from protecting their eggs or chicks. These include fights and nest reliefs. Birds can be knocked from nests in fights, even if they are not the combatants. If nest reliefs are too slow, then eggs or small chicks are exposed for longer than they should be.
- Any eggs or chicks (that have not learned nest location; ch. 6), if pushed from the nest, are not retrieved by parents. These eggs and chicks become skua food.
- Penguins incubating eggs face into the wind, and those brooding chicks keep their backs to the wind.[31,39] If the wind is strong enough that all members of a subcolony face the same way, then their group vigilance is reduced. A skua can more easily approach from behind or remain unnoticed close by until an opportunity arises to make off with an egg or chick. However, if winds are this strong, then the adults probably are incubating or brooding much closer than otherwise. In fact, if it is calm and warm, chicks might well be more exposed because the parents stand by the nest panting.

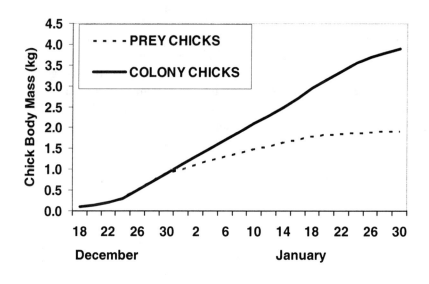
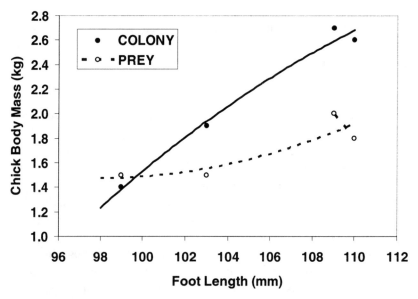

Figure 7.1 The size of chicks taken by South Polar skuas compared with the average size of chicks in the colony. A) Penguin chick mass as a function of date on Ross Island. B) Condition of the average chick in the colony compared with those taken as prey, by ratio of body mass to size (measured by foot length). *Figure redrawn from data contained in Young (p. 320).*[41]

- The penguins' advantage of nesting synchronously breaks down at the start of the postguard stage (beginning of crèche stage). The first chicks left unattended are vulnerable to skuas, depending on size and location in the subcolony, because they are pushed away by other parents (ch. 6).
- Feeding chases take chicks away from the protection of the crèche and of subadults within the subcolony (ch. 6). Underweight chicks are especially vulnerable; they are light enough to pull away and are too weak to defend themselves (fig. 7.1).

FACTORS THAT CONSTRAIN SKUAS

Young[41] also discussed the characteristics of penguin breeding biology that prevent skuas from taking penguin eggs and chicks:

- The tight synchrony of penguin breeding, caused by the short summer season, dictates when the penguin resource is available (fig. 7.2). The skua nesting cycle lags that of the penguin by about a month,[30,41] with the result that the skua runs out of penguin food before its cycle is completed. Also, only penguin eggs are available during the skua's courtship and laying periods. Important in the skua's pair formation and egg formation processes is feeding of the

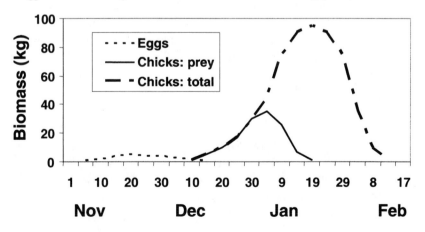

Figure 7.2 Conceptual model showing the total mass of eggs and chicks in a hypothetical colony of 25,000 breeding penguins, and the mass available to skuas as a function of egg and chick size. *Figure redrawn from data contained in Young (p. 76).*[41]

female by the male (courtship feeding). Rarely does a male bring a penguin egg to his mate at the nest; almost always it is a fish (but if the female is near the penguins, the male will let the female eat the egg contents first). Finally, the penguin cycle is so synchronous that only one type of penguin food is available at a given time: first eggs, then young chicks, then young postguard chicks. Because of their size, mature chicks are mostly off limits to South Polar skuas.

- The colonial nesting habits of penguins constrain skuas' movements and access to penguin eggs and chicks. An Adélie penguin is too physically imposing to a skua, and because the penguins nest so closely together, it is dangerous for a skua to move between penguin nests. Also, the penguins alert one another when a skua alights nearby. The first thing heard are sneezes as the penguins are aroused; the chorus of sneezes bring other birds to alert.[1] If the skua is very close, then a growl alerts all penguins in the vicinity.
- The penguins, particularly older breeders, are very protective of eggs and chicks. The nest bowl helps the penguin protect its contents to some degree. The skua's success also depends on the number of eggs or chicks and the size of chicks. If there are two eggs or chicks, often one may stick out more than the other. A skua can grab it more easily, especially if the chicks are large enough to stick out.
- Skuas have a much more difficult time if many penguin adults are present. Prebreeding adult penguins, which arrive and investigate the nesting process while chicks are present, are free to chase skuas. In fact, early naturalists called these birds the guardians of the crèche. By chasing skuas, these birds are not acting out of care for the well-being of the chicks, however; they are acting on their own dislike for skuas.[30] Nevertheless, their presence is a major deterrent. Without the usual numbers of such penguins (which happens when food is less available or farther away), the skuas become bolder and can even take larger chicks than usual. Also, fights between reoccupying birds and young adults can expose chicks to skuas.

Finally, Young[41] discussed constraints to the skuas imposed by the size and strength of Adélie penguins:

- As mentioned earlier, skuas are very wary of approaching a penguin that is aware of its presence. An adult skua weighs 0.9–1.5 kilograms, with females larger than males.[4] Like those of all aerial birds, a skua's bones are hollow and therefore brittle. An adult Adélie penguin, on the other hand, weighs 3.8–6.0 kilograms depending on sex and time of year; its bones are dense, and its densely feathered skin is very tough. A whack from its flippers can easily break the skua's bones. It is unlikely that a skua could do any damage to an adult Adélie penguin.

- A South Polar skua can lift 450 grams and drag 600 grams.[41] In winds stronger than 10 knots, which provide more lift, a skua can carry 550 grams and pull 700 grams. Therefore a skua would have difficulty with a normally fed chick older than ten days (fig. 7.1). Smaller chicks are much more vulnerable, especially soon after hatching, because they can be hooked by the skua's bill and easily carried away. In fact, a small chick can be stolen far more easily than an egg. A skua struggling to drag a large chick from the subcolony edge is very susceptible to attack by prebreeding adult penguins. Skuas must immobilize chicks quickly if they are to succeed in killing them. Chicks that are 2,000 grams or larger (i.e., three to four weeks and older; fig. 7.1) are invulnerable to skuas.
- Once the penguin feeding chases begin, chicks must return to the subcolony after being fed. The running speed of chicks when twenty to twenty-five days of age is 0.9–1.2 meters per second, and by forty days it has increased to the speed of adults, 2.2 meters per second (8 kilometers per hour). Therefore the chicks' strength and coordination increase rapidly, allowing them to return quickly to the protection of the subcolony and making it more difficult for the skua to pull them away. Only recently crèched chicks are vulnerable because of their smaller size and lesser strength and speed. Older chicks are beyond the capabilities of the skuas (fig. 7.2).

Almost all eggs and chicks lost by Adélie penguins are eaten by skuas,[41] whether they obtain them by fortune or predation. The proportion of eggs consumed by skuas averages about 31 percent of those laid, ranging from 12 to 63 percent depending on colony and year. This is a great many eggs in a large penguin colony. As noted by Sladen[30] and others, the territories of some skua pairs become densely littered with penguin egg shells. Possibly, only 1–2 percent of eggs lost would not be lost were it not for skuas taking them; the remaining 98–99 percent would be lost to other causes.[17] The 1–2 percent would constitute true predation of eggs.

The taking of eggs by skuas occurs mainly during two short periods.[17] The first peak occurs at penguin egg laying, when eggs are lost by penguins that do not protect them closely. As noted in chapter 5, these eggs include those of young, first-time breeders or from nests about to be deserted by birds that fasted too long (e.g., reversed incubation routines). Most of these eggs are lost not long after they are laid, so there is a peak in egg snatching by skuas at the latter part of the peak of egg laying, when young Adélie penguins are laying. This is also a time—before the skuas themselves have laid eggs—that a pair of skuas can work together, one to distract and the other to snatch. Later, when one skua parent must incubate or brood small chicks, the breeding skuas must work alone. Then it is by stealth and great patience, waiting long periods, that skuas find opportunities to capture an untended egg or chick. The second peak, involving half of all eggs taken, involves eggs rejected by penguin parents after the first egg of the clutch hatches. Therefore this second peak in egg taking occurs at the peak of penguin egg hatching. This is a transitional period when penguin parents quickly become more attentive to a new chick than to the other egg. Second eggs laid later than usual often fall to this fate. These second eggs often are exposed, which makes them more vulnerable. Really late eggs are ignored by skuas because they probably are rotten.

PREDATION OR SCAVENGING?

True predation occurs when a skua takes and kills chicks able to survive away from the nest. As noted earlier, most chicks are taken during transitional periods. The first is at peak of hatching, when it is much easier for a skua to snag a small chick than to grab an egg. Also, chicks are much more nutritious than eggs.[41] The second peak in the taking of chicks occurs early in the crèching period, when first-crèched chicks are rejected by other parents and tend to be small enough for a skua to manage (especially if underfed). Both of these peaks occur in the first third of the penguin chick growth (fig. 7.2).

The results of exclosure experiments are especially interesting.[41] Wires were strung a meter above the penguins' heads in sections of subcolonies. The wires inhibited overflights by skuas. Results indicated appreciable egg and chick mortality independent of skua disturbance or predation. In one group of penguins, nesting success to guard stage differed only slightly between protected and unprotected birds (82 percent versus 87 percent, unprotected higher); in another group the difference was 86 percent versus 47 percent. Results also varied by year. There was no effect between protected and unprotected nests in 1968–69 (year of extensive ice) compared with the previous year (ice extent normal). In other words, in the year when the penguins' fat reserves were low, they lost eggs regardless of protection and mostly because of desertion of nests before mates could return.

Do skuas with penguins in their territories do better? Counts of the number of active penguin nests within the territories of skuas nesting at the edge of penguin colonies vary from 700–3,000 at Crozier[20] to 730–3,190 at Cape Hallett.[33] At Cape Bird, holding a territory containing penguin nests did not appear to confer an advantage to a skua pair.[41] That is, there was no effect on nesting success when comparing skuas with or without penguins in their territories. On the other hand, at Cape Hallett the skuas having territories that contained penguins did better than those that did not.[33] However, their enhanced success did not result from the penguin resource. Rather, skuas defending territories holding penguins tended to have much larger territories than other skuas. Therefore the penguin-defending skuas suffered less nest predation from neighboring skuas.

It has also been observed that skuas that defend territories containing penguins tend to lay their own eggs three to six days earlier than skuas without penguins.[33,34,41] It is possible, but not known for certain, that this effect arises because skuas with penguins begin to defend territory earlier than other skuas (to protect the penguin resource from intruders), and this speeds their nesting process. It is not the nutrition from the penguin resource itself that is responsible. The consensus is that earlier laying by skuas having penguins in their territories is an advantage because penguin chicks are available for a longer portion of the skua chick period. In the studies conducted thus far, no increase in skua nesting success among skuas with penguins has been detected. That is, no evidence has yet surfaced to indicate a positive effect of the penguins as a food resource. On the other hand, in many seabird species, chicks that fledge late often suffer increased mortality. Therefore nesting early, which would lead to early fledging of chicks, ultimately may increase nesting success by enhancing the postfledging survival of the skua chicks.

Are penguins that nest within a skua territory more successful or less successful than those that do not nest within a skua territory? At a small colony, such as Cape Royds (4,000 pairs, 75 × 150 meters), almost all penguins nest within the territory of one of several skua pairs that nest around the colony's periphery. In contrast, only the penguins that nest in the outer 40–50 meters of a large colony, such as Cape Crozier (120,000 pairs, 1 × 2 kilometers), nest within the territory of a skua. The majority of penguins in that case nest away from the colony edge. It appears that penguins nesting within a skua territory fare no better or worse than those that do not.[20,33] It depends on whether a given skua pair is particularly aggressive in obtaining penguin eggs and chicks; most skuas are not.[41]

To summarize the skua-penguin relationship, Young said,

> The overall conclusion . . . is that over much of the colony skuas do not have a major impact on penguin breeding. For most [subcolonies] it is exceeded by the activities of the penguins themselves. That is, it is hard to demonstrate . . . that the skuas are in general strong predators on the penguins. The evidence suggests on the contrary that most of their food is taken by scavenging, and that most of this comes not from their activities disturbing the penguins but from the effect of penguin and environmental factors. In some places on the colony, however, there was undeniable predator impact with very high egg and chick losses. (p. 407)[41]

Chapter 8

Demography

The number of chicks produced per breeding pair is known in demographic analyses as fecundity. As we have seen in earlier chapters, the number of chicks fledged is affected strongly by factors at play in the colony, such as parental age, nest location, and timing of breeding. The number of chicks fledged is also strongly affected by parents' success at obtaining food. Along with fecundity, other important demographic variables are breeding propensity, or the proportion of the population that attempts to breed among those physiologically capable of doing so; age at first breeding, which indirectly defines physiological capability; and annual survival. All of these variables vary according to the age of the cohort. Together, they determine whether a population or colony will grow or decline.

Unlike fecundity, mortality of adult Adélie penguins is related mainly to activities at sea. If they can cope, they survive; if not, they die. In contrast to our detailed knowledge of factors affecting colony activities, we know much less about those affecting survival because the pelagic lives of these birds are far less accessible to us. Because Adélie penguins spend only 10 percent of their lives on land,[2] it would be absurd to say that their marine activities are only a minor part of their existence or that understanding their breeding biology is the key to knowing these creatures.

Chapters 2 and 7 summarize what is currently known about the marine ecology of Adélie penguins; as noted therein, much more remains to be learned. What little direct and indirect knowledge we have indicates that age and experience relate importantly to the behavior of Adélie penguins in their marine habitat. The latter subject is addressed more fully at the end of this chapter.

The demography of Adélie penguins in the Cape Crozier population was initially analyzed by David Ainley and Douglas DeMaster.[3] For individuals six years and older, annual survivorship varied from 58 to 89 percent, a range that includes the estimates of adult mortality for Adélie penguins at capes Hallett, Bird, and Royds.[45,53,56] However, at least for the Cape Crozier population, the figure probably is underestimated. Annual survivorship has been determined to be about 87 percent in the longer-lived royal and yellow-eyed penguins,[3,13] about 89 percent in the African penguin,[17,49] and about 75 percent in the shorter-lived little blue penguin.[20]

Survivorship of Adélie penguins in the Cape Crozier population was lower among younger breeders and among breeders (61 percent) than among nonbreeders (78 percent) regardless of age.[3] Survivorship was especially low in first-time breeding three-year-old females (25 percent) and four-year-old males (36 percent). Such patterns are not unique to Adélie penguins. King penguins that raise chicks to fledging have higher mortality than those that do not.[55] Northern fulmars that commence breeding at a young age have higher mortality in their first breeding year than those that begin breeding at older ages.[22,41] The same is true for short-tailed shearwaters[61] and kittiwakes.[15,62]

Only a few Adélie penguins reach sixteen years of age, and those that do so are individuals that first bred at a late age (six years or older; table 8.1). Subsequently, they tended to be nonbreeders or failed breeders.[1] Why do young Adélie penguins even attempt to breed when their chances of doing so successfully and surviving are so low? Obviously they find it difficult to cope simultaneously with demands of the sea as well as the breeding colony.

The following is a reanalysis of the demography of the Cape Crozier population during the 1960s and early 1970s. It was accomplished with much help from Doug DeMaster, and except for some of the discussion of more recent research, what is presented is fairly close to what appeared in my earlier monograph on Adélie penguin breeding biology.[4] The information presented does not necessarily apply to other Adélie penguin

Table 8.1 Percentage attaining various minimum ages at death by age at first breeding for male Adélie penguins, Cape Crozier, 1965–1974.

	Minimum Age at Death			
Age First Bred (yr)	*5–7*	*8–10*	*11–13+*	*n*
4	6%	77%	17%	13
5	9	34	57	35
6	5	3	92	37
7–8			100	16

Age at death assumes the bird lived only through the last year seen.

populations. It is important to note several facts that have become apparent only with the perspective of time. First, the Cape Crozier colony was decreasing in size during the period of study, as were all Ross Island colonies at the time. The reason for the decrease appears to be related to a period of especially severe sea-ice conditions. If the colony population had been stable or increasing, then it would be reasonable to expect that some of the data discussed in this chapter would be different. Second, the analysis of emigration and immigration rates (a measure of philopatry, or the tendency of individuals to return to their natal colony) is biased toward Cape Crozier during the period of population decrease. As noted in chapter 3, Cape Crozier is the largest colony of an isolated cluster in which all other colonies are much smaller. Banding of chicks, which allowed the study of philopatry, occurred only at Cape Crozier. Thus the measure of philopatry is from the perspective of the largest colony in the group. Given that Cape Crozier was closest to the Ross Sea Polynya and, accordingly, more dispersed pack ice, it is distinctly possible that emigration from Crozier (to the more severe conditions of outlying colonies) was weak. Conversely, immigration from the smaller, outlying (and iced-in) colonies to Cape Crozier perhaps was much stronger. Moreover, if the large Cape Crozier colony had been expanding, stronger emigration from it might be expected to have been even more prevalent. The opposite patterns might be expected if colonies were expanding in size (which they were in the 1980s and 1990s; chapter 9) and more dispersed pack ice was available to the outlying colonies (which it was in the later years).

■ Why Three-Year-Olds Risk Breeding

It appears that birds that first breed at a young age tend to breed successfully in more subsequent seasons than birds that breed first at older ages (table 8.2). Thus a larger proportion of birds that first bred as three- or four-year-olds during the 1965 to 1969 period at Cape Crozier bred successfully in a larger proportion of subsequent seasons. Birds breeding successfully in their sixth (or later) season were observed doing so during 1974–75 and 1975–76 and must have first bred early in the 1965–1969 period. This would actually decrease their chances of surviving in the later years merely by the accumulation of time (and chance). The oldest known-age birds in 1965 were four years old, and in 1969 the oldest were eight years old. An increase in breeding success during an entire breeding lifetime provides the selective advantage of breeding among the youngest mature Adélie penguins despite high mortality. In other words, birds that breed at an early age tend to be better breeders because of some unknown behavioral or physiological characteristics. The result is that this minority of individuals contributes far more offspring to the population than do the majority of adults, a pattern that has been found among other seabirds.[11,20,37]

These trends are confirmed by the fact that a higher proportion of females first breeding at three years of age attempt breeding in more subsequent years than birds first breeding at older ages. Note that reference here is to breeding *attempts* rather than successful breeding, as in the preceding paragraph. The absolute number of breeding years is compared for females in table 8.3, but to increase sample size the maximum possible number of breeding years is compared in table 8.4. In both samples, all

Table 8.2 Number and percentage of Adélie penguins breeding successfully in a give number of seasons by the age at which they first bred.

	Number of Seasons Successful			
Age of First Breeding (yr)	2	3	6+	n
3	6 (38%)	3 (19%)	7 (44%)	16
4	20 (40)	9 (18)	21 (42)	50
5	19 (40)	10 (21)	18 (38)	47
6	8 (47)	4 (24)	5 (29)	17

Table 8.3 Percentage of female Adélie penguins breeding for a known number of years by age at first breeding, Cape Crozier, 1965–1975.

	Years Known to Have Bred		
Age First Bred (yr)	2	>2*	n
3	43%	57%	7
4	63	37	19
5+	60	40	5

Table 8.4 Percentage of Adélie penguins breeding for the maximum possible number of years by age at first breeding, 1965 and 1975.

	Maximum Years Bred		
Age First Bred (yr)	2–6	7–9*	n
Female			
3	45%	55%	20
4	56	44	63
5	53	47	45
6–7	59	41	22
Male			
4–5	58	42	48
6	67	33	36
7–8	75	25	16

*Comparing the proportion for youngest breeders with the weighted value for older ones (females, 0.44; male, 0.30) results in the following values of t: females 0.918 ($P > .05$), males 6.992 ($P < .05$).

birds had to be three years of age by the 1968–69 season. An insufficient sample of males was available for inclusion in table 8.3, but table 8.4 shows results for males similar to those for females. These tables compare the proportion of birds breeding for the largest number of subsequent years among those first breeding at the youngest and oldest ages. A significant difference is evident only for males, possibly because sample sizes are more adequate for them (in that males fail to breed much more often than females). Comparing what they called "continuous" with "intermittent" breeders, Richard Wooller and John Coulson[62] found a higher proportion of continuous breeders among kittiwakes that first bred at older ages than among those that did so at younger ages. However, they did not make their comparison on the basis of the number of nonbreeding years, as is the case for Adélie penguins. Almost all male Adélie penguins, and perhaps a higher proportion of females than in kittiwakes, are intermittent breeders. This is because Adélie penguins must find new mates much more frequently because of predation; this leads to a tendency to skip breeding during the re-pairing process.[53]

The mortality rate of Adélie penguin breeders declines with age but is still higher than that of nonbreeders; the mortality of nonbreeders remains unchanged with increased age except for a sharp rise among the very oldest birds.[3,4] Those results and the ones presented earlier indicate that if adult Adélie penguins are to replace themselves, they must participate in a race between producing offspring and surviving long enough to do so. Clearly, there would be little selective advantage for Adélie penguins younger than three years to breed. This conclusion assumes that the infertility rate among two-year-old Adélie penguins would be as high as it is in two-year-old yellow-eyed penguins, where an incredible 65 percent of eggs incubated to term fail to hatch (and probably were infertile).[46] That conclusion also takes into account the very high mortality among the youngest breeding Adélie penguins.

■ Fecundity

The interplay between productivity and mortality determines whether and to what degree a population is increasing, decreasing, or unchanging. To determine stability, or growth (λ), estimates for two factors are needed: fledging rates for parents of known age and survival rates for these same age groups. In practice, however, demographic studies must deal with two populations, a tagged and an untagged one. It then becomes necessary to determine how well the tagged population represents the other in calculations of these values. In the Cape Crozier study, differences in these two populations did occur.

The overall productivity of known-age Adélie penguins in the Cape

Table 8.5 Average number of chicks fledged by parent's age among Adélie penguin adults attempting to breed (i.e., eggs observed), Cape Crozier, 1965–1974.

	Male Parent		Female Parent		
Age (yr)	Chicks Fledged	n	Chicks Fledged	n	Weighted Averag
3	0.0	0	0.3 ± 0.5	50	0.3 ± 0.5
4	0.2 ± 0.5	34	0.4 ± 0.6	133	0.4 ± 0.6
5	0.4 ± 0.6	174	0.6 ± 0.7	228	0.5 ± 0.7
6	0.8 ± 0.7	216	0.9 ± 0.7	221	0.9 ± 0.7
7+	1.0 ± 0.8	303	1.0 ± 0.7	262	1.0 ± 0.8
Total	0.7 ± 0.7	727	0.7 ± 0.7	894	0.7 ± 0.7

Sample size refers to the number of breeding pairs observed.

Table 8.6 Chick production in a breeding population of Adélie penguins, Cape Crozier, 1968–1975.

Age (yr)	Cohort n*	Adult Survival, %[†]	Corrected Cohort, n	Proportion Adults Breeding[‡]	Breeding Population, Pairs	Fledglings Produced[§]	Total Fledglings
3	7,400	0.212	1,569	0.09	141	0.3	42.3
4	6,880	0.148	1,018	0.30	305	0.4	122.0
5	10,400	0.104	1,082	0.52	563	0.5	281.5
6	14,380	0.072	1,035	0.73	756	0.9	680.4
7+	61,160	0.025	1,520	0.85	1,292	1.0	1,292.0
Total					3,057		2,418.2

*From Ainley and DeMaster (tables 1, 7).[3]
[†]From Ainley and DeMaster (figure 4).[3]
[‡]From table 5.3.
[§]From table 8.5.

Crozier population during the mid-1960s to mid-1970s is summarized in table 8.5. Combining these data with others (table 8.6) indicates that 0.8 chicks (2,418 chicks per 3,057 pairs) were fledged per breeding pair. This figure agrees with the average for all studies in which breeding success has been estimated: 0.9 ± 0.1 (table 5.10, without the inexplicably

Table 8.7 Data on fecundity (mx) in Adélie penguins, Cape Crozier, 1967–1974.

Age (yr)	Fledglings per Breeder	Percentage Breeding	Female Fledglings per Bird
3	0.3	17	0.026
4	0.4	49	0.098
5	0.5	68	0.204
6	0.9	82	0.369
7+	1.0	90	0.450

high estimates at Magnetic Island,[58] as discussed in chapter 5). This similarity supports the idea that the productivity data on banded birds adequately described values for unbanded ones.

The data in table 8.6 were then used to determine the number of female chicks produced by all females (including nonbreeders) in an age group. This is the conventional measure of fecundity (m_x) in demographic analysis (table 8.7). Combined with survivorship data,[3] the resultant net reproductive rate (R_0; see below) became 0.116. This low figure implies that the existence of Adélie penguins at Cape Crozier is problematic or that something was wrong with the data. Subsequently, it has become clear that the population was indeed declining during the study period, as discussed earlier. However, the decline was not as abrupt as indicated by the R_0 calculated. The problem probably resided in the survivorship data.

■ Survivorship and Age Structure

The Crozier data were reviewed to identify any biases not previously apparent in the derivation of the survivorship values, the probability of sighting a banded bird among the thousands of birds present, and the assessment of band loss.[3] Also, there is the question of whether banding itself caused mortality. In the Crozier study, flipper bands were used.[50,51] The following points resulted from this inquiry.

First, the probability of sighting a banded bird differed, as would be expected, between the part of the colony visited by researchers daily and the part visited less frequently. In the former, for birds five years of age and older during the 1974–75 and 1975–76 seasons, probability of sighting averaged 0.977, compared to 0.888 in the latter ($P < .05$; t test). Therefore restricting analysis to data collected from intensively studied areas is justifiable,[3] and that procedure was followed.

The probability of sighting also varied according to the age of the birds. These probabilities were calculated using the formula

$$l = \frac{n_{i+1} s^{i+1} p_{i+1}}{n_i s^i p_i},$$

where n = number of birds seen of an age group in year i, s = age-specific survivorship, and p = probability of sighting. The formula reduced to

$$p_i = \frac{s p_{i+1} n_i}{n_{i+1}}.$$

For the purposes of estimation it was assumed that $p_2 \neq p_3$, $p_3 \neq p_4$, $p_4 = p_A$ and that s was constant for all ages greater than one year. The following estimates were made: p_A equaled 0.98 (table 8.8); p_2 and p_3 equaled 0.215 and 0.902, respectively (table 8.9); and s equaled 0.894.

Second, banding or, more properly, the bands themselves apparently caused some mortality. For known-age birds rebanded as four- through

Table 8.8 Estimation of the probability of sighting [p(sight$_1$)] for adult Adélie penguins, Cape Crozier, 1974–1976.

Age (i) (yr)	n_i	n_{i+1}	p(sight)
5	348	360	0.967
6	435	442	0.984
7	316	325	0.972
8	78	78	1.000
9	50	51	0.980
10	44	46	0.957
11	30	30	1.000
12	24	24	1.000
All*	1,325	1,356	0.982

*Weighted mean = 0.977.

Table 8.9 Estimated probability of sighting [(sight$_1$)] for young Adélie penguins, Cape Crozier, 1967–1969.

Age (i) (yr)	n_i	n_{i+1}	p(sight)
2	206	773	0.215
3	951	924	0.902

s and p_A equal 0.894 and 0.98, respectively.

Table 8.10 Summary of data on annual survivorship of Adélie penguins as a function of the number of years after rebanding, Cape Crozier, 1965–1969.

	Years After Rebanding			
	1	2	3	4
No. rebanded birds	364	309	222	139
No. survived	196	231	166	100
Sx weighted	0.538	0.748	0.748	0.719
Range in values	0.45–0.61	0.66–0.77	0.72–0.80	0.69–0.80
No. cohorts analyzed	7	6	4	3

seven-year-olds during the 1964–65 to 1967–68 seasons, their survival to the first year after rebanding differed markedly from their survival from first to second, second to third, and third to fourth years after rebanding (table 8.10). The survival of birds with new bands was 72 percent of the survival rate of birds with bands affixed longer than one year.

Based on observations of captive Adélie penguins at SeaWorld (Scott Dreishman, personal communication, 1980), mortality of penguins with flipper bands may result from complications arising when the wing swells during molt and the band constricts blood flow. The increased mortality (i.e., 28 percent more birds die than should) probably would occur only once: during the first molt, which happens when an Adélie penguin is thirteen to fourteen months old. Whether an additional but lower band-induced mortality affected survivorship in older birds is not known. For banded and unbanded birds, S_0 was estimated as follows:

$$l_{x(2)} = \frac{n_2}{[1 - p(\text{band loss})_2]\, p(\text{sight}_2)\, n_0},$$

where $p(\text{band loss})_2$ equals the probability of losing a band in two years; $p(\text{sight}_2)$ equals the probability of seeing a two-year-old at the colony, given that it is alive; n_2 is the number of two-year-old penguins seen; and n_0 is the number of chicks originally banded. In 1965–66, 3,800 chicks were banded, and two years later 153 two-year-olds were observed at Cape Crozier. Band loss in two years was estimated[3] to be 0.01, and $p(\text{sight}_2)$ was estimated to equal 0.215. Therefore $l_{x(2)}$ equaled 0.189. Let

$$l_{x(2)} = (fs_0)s_i$$

where f equals the ratio of the annual survivorship of a banded individual to that of an unbanded individual, and s_i equals the annual survivorship of an individual of age i to $i+1$. Assuming that $s_0 = s_i$, that f equaled 0.72, and that $l_{x(s)}$ equaled 0.189, then s_0 equaled 0.513. This constituted the probability that an unbanded fledged chick survived one year. The survival rate for a banded chick equaled 0.369 (0.513×0.72; s_i for a banded chick was assumed to equal s_i for an unbanded chick, or 0.513).

Third, the data presented previously on band loss[3] could not be improved. Still questionable was whether these data were adequate because band loss values seemed a little low compared to those of other studies in which aluminum bands were used.[28] (Unfortunately, aluminum is quickly softened by chemical reaction with seawater. Today, seabird researchers use stainless steel bands.) In addition to the loss itself, the degree of independence between sightings of bands and sightings of web punches was questionable (the toe webs of banded chicks were also clipped in combinations unique to each cohort.)[4] Truly naïve observers (i.e., ones having no idea where banded birds resided) were not available to search for web punches (only upon sighting a bird with a punched web was the observer supposed to look for its band).[3] Furthermore, it was not known whether some punched webs healed over (probably not) or whether punched webs induced any mortality (again, probably not).

Annual survival estimates for 1968–69 and 1969–70[3] were combined with those for 1967–68[4] to determine average survivorship estimates for four- through seven-year-olds (table 8.11). These were improvements over minimum survival estimates.[54] The degree of annual variation in these figures has proved to be similar to those for breeding Adélie penguins at Cape Bird[53] during the same years. Data from 1974–75 and 1975–76 were not used because they were only two consecutive years and because an unusually high disappearance rate (mortality) occurred

Table 8.11 Estimated adult survivorship of Adélie penguins, Cape Crozier 1967–1969.

Age (yr)	Year	Y_i	Y_{i+1}	Banding Mortality	Band Loss	Emigration	Total	S_x
4	1967–68	232	222	0	1	2	225	0.970
5	1967–68	129	109	12	2	2	125	0.969
	1968–69	218	152	13		82	175	0.803
6	1967–68	35	30	2	1	1	34	0.971
	1968–69	109	76	3	5	2	86	0.789
7	1968–69	30	24	2	1	1	28	0.933
All	1967–68	396					384	0.970
All	1968–69	357					289	0.810
All	1961–69	753					673	0.894

among breeding birds during 1974–75 (i.e., they failed to return from feeding at sea). For example, looking just at the incubation period, 10.2 percent (n = 254) of breeding birds disappeared in 1974–75, compared with 5.8 percent (n = 138) in 1968–69 (P < .05, t test).

Using a 20 × 20 Leslie model[33] with age extended to nineteen years, estimates S_0, S_1, and S_A were combined with fecundity (table 8.7) to calculate population growth (λ). Actually, in 1982, I found a twenty-year-old Adélie penguin, which finally justified extending the model beyond the ages available in the survivorship data. This is the oldest Adélie penguin ever sighted in the wild. In the banded population (S_0 = 0.369, S_1 = 0.513, and S_A = 0.894), λ = 0.96. The banded population thus was declining at a 3 percent greater annual rate than the unbanded one. Using these values of λ, the age structure of the unbanded and banded populations was calculated (table 8.12). Age structure in both was similar, with roughly 20–25 percent of each being fledged chicks and 50 percent being birds at least three years old. This indicates high productivity, which is consistent with data on breeding biology reviewed in chapters 5 and 6 and high mortality in older age groups. The latter attribute is consistent with the idea that Adélie penguins are unique among seabirds by not being at the top of the food web.

A DECLINING POPULATION AT CAPE CROZIER IN THE 1960S AND 1970S

The sum of $l_x m_x$ (or R_0, the net reproductive rate) in the unbanded Cape Crozier population was 0.616, and in the banded population it was 0.443. The value originally calculated with the survivorship data in Ainley and DeMaster[3] was 0.116. The 3 percentage point difference in the two (values calculated earlier further indicated that both the banded and unbanded populations at Cape Crozier were declining during the 1960s, one slightly faster than the other. Taking into consideration all data reviewed in this book, it became apparent that the decline was real and was not an artifact of inadequacies in the data. With $\lambda = 0.96$, the population should have declined by about 34 percent over ten years, a figure in accord with the trend illustrated in figure 8.1. In retrospect, we now know that the Adélie penguin colony at Cape Crozier and other colonies on Ross Island (and probably the entire Ross Sea) were indeed declining in size during the 1960s and 1970s.

Table 8.12 Age structure in a banded ($\lambda = 0.9333$) and an unbanded ($\lambda = 0.959$) population of Adélie penguins, Cape Crozier, 1961–1976.

	Banded Population				Unbanded Population			
Age (X) (yr)	S_x	l_x	$l_x(\lambda^{-x})$	$C_o l_x(\lambda^{-x})$	S_x	l_x	$l_x(\lambda^{-x})$	$C_o l_x(\lambda^{-x})$
0	0.369	1.000	1.000	0.240	0.513	1.000	1.000	1.219
1	0.513	0.369	0.395	0.095	0.513	0.513	0.535	0.117
2	0.894	0.189	0.217	0.052	0.894	0.263	0.286	0.063
3	0.894	0.169	0.208	0.050	0.894	0.235	0.266	0.058
4	0.894	0.151	0.199	0.048	0.894	0.210	0.248	0.054
5	0.894	0.135	0.191	0.046	0.894	0.188	0.232	0.051
6	0.894	0.121	0.183	0.044	0.894	0.168	0.216	0.047
7	0.894	0.108	0.175	0.042	0.894	0.150	0.201	0.044
8	0.894	0.097	0.169	0.040	0.894	0.134	0.187	0.041
9	0.894	0.086	0.160	0.038	0.894	0.120	0.175	0.038
10	0.894	0.077	0.154	0.037	0.894	0.107	0.163	0.036
11	0.894	0.069	0.147	0.035	0.894	0.096	0.152	0.033
12	0.894	0.062	0.142	0.034	0.894	0.086	0.142	0.031
13	0.894	0.055	0.135	0.032	0.894	0.077	0.133	0.029
14	0.894	0.049	0.129	0.031	0.894	0.069	0.124	0.027
15	0.894	0.044	0.124	0.030	0.894	0.061	0.114	0.025
16	0.894	0.039	0.118	0.028	0.894	0.055	0.107	0.023
17	0.894	0.035	0.113	0.027	0.894	0.049	0.100	0.022
18	0.894	0.032	0.111	0.027	0.894	0.044	0.093	0.020
19	0.894	0.028	0.104	0.025	0.894	0.039	0.086	0.019
Totals			4.174	1.001			4.560	1.997

S = age-specific survivorship; l = proportion alive of a given age class (x); $C_o l_x(\lambda^{-x})$ is the proportion of that age class in the population.

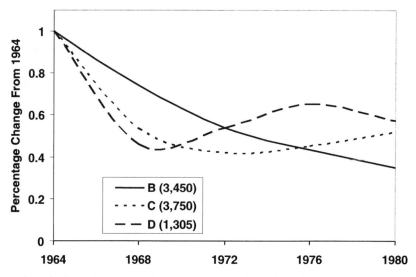

Figure 8.1 Trends in the size of groups of subcolonies at Cape Crozier. These were the subcolonies in which chick banding occurred; the initial size, in pairs, is given in parentheses. Research concentrated in areas B and C; area D was used as a control to study possible effects of disturbance by researchers and therefore was rarely visited.[4]

To summarize the major points thus far in this chapter, we found the following: Flipper bands induced mortality during the first year after banding but not thereafter; probability of sighting was age specific; survivorship rates determined earlier[3] are in the lower range of annual variation; age structure confirmed high productivity, with 20–25 percent of individuals in the population being fledged young and almost 60 percent being birds in the breeding population; and the banded population of Adélie penguins at Cape Crozier was declining 3 percent more rapidly than the unbanded one, which was declining at about 4 percent per annum during the 1960s and early 1970s (λ banded = 0.93, λ unbanded = 0.96). The overall significance of these findings in relation to those of preceding chapters is a complex matter.

■ Why Is Breeding Deferred?

Chapter 4 closed with a discussion of the adaptive advantages of delaying visits to the colony until the penguin is older than one year (two to five years for most individuals) and, when making first visits, to wait until late in the summer. The conclusion was that a few years pass before Adélie penguins can feed proficiently enough to visit the colony and survive the necessary fasting. Also noted was the fact that young birds arrive late in the summer, when sea-ice conditions are more favorable. It is now appropriate to continue that discussion by confronting the more basic questions of why maturity is deferred in the Adélie penguin and how deferral contributes to population regulation. The answers are not simple because physiological, ecological, behavioral, and demographic factors are involved. Proximate factors that might limit breeding include limitations on nesting space, mates, or food. The ultimate factors involve delay in maturation.

SPACE LIMITATIONS

Usually most adults prevented from breeding by lack of territory are young, so in effect they are prevented from breeding beyond the age at which they reach physiological maturity. Lack of a territory usually limits breeding in cavity-nesting species.[12,57] The possibility that this happens among seabirds has been demonstrated directly in only one population, that of the Cassin's auklet (*Ptychoramphus aleuticus*) on the Farallon Islands, California, where the number of nest cavities was limited, at least through the 1970s.[35] In the case of the cavity-nesting African penguin, it is possible that at least before the present demise of the population caused by oil pollution, guano harvesting, fishery overharvest, and perhaps other anthropogenic factors,[49] lack of nesting territory could also have limited breeding population size. For that species, offshore islands suitable for nesting are few, competition for nesting space is intense,[16] variation in day length is minimal, and food was once available for much of the year.[18] In response to photoperiod and food, African penguins nest throughout the year, and individual pairs can produce two or even three broods in succession.[44] Such a situation is likely to prevent certain individuals from nesting for lack of space (assuming preexploitation population size).

Limitations on breeding caused by lack of a territory among young individuals of open-nesting seabirds have been inferred in at least one study in which adult gulls were culled from the breeding population (as a management tool).[14] In that situation, the colony definitely was overpopulated. However, such a situation is not likely for open-nesting seabirds such as the Adélie penguin. At most colonies, including the huge one at Cape Crozier, every male Adélie penguin who arrives during the spring occupation period can find a territory. Large expanses of seemingly suitable nesting habitat go unclaimed.

MATING LIMITATIONS

For male Adélie penguins, availability of mates is a limiting factor, and the competition probably does prevent some from nesting. By six years of age, 45 percent of males and 9 percent of females at Cape Crozier had still not bred (table 5.2), and it was not until about eight years that the incidence of breeding in males reached the peak level of females. For females, peak breeding incidence was established at six years of age, with 80–90 percent breeding (table 5.3). These statistics indicate a shortage of females to pair with territorial males.

Many of the older nonbreeding Adélie penguins and almost all of the few older nonbreeding females acquire mates but for some reason fail to produce eggs in a given season. Thus the shortage of females is much less than it first appears. In birds eight years and older, 17 percent of males do not breed, compared with 11 percent of females ($t = 2.36$, $P < .05$; table 5.3). The difference in these percentages is the actual proportion of males in the adult population that do not breed because of the shortfall in female numbers. By this reasoning, a 6 percent surplus existed among the males of breeding age at Cape Crozier. These males lost the competition for mates because of inadequate social behavior. Even without a shortfall in availability of females, they still would have failed to breed or even establish a pair bond.[1] They were among the oldest of all first-time breeders and were nonbreeders during many of their later seasons at Cape Crozier. Most were unable to reproduce, although they were among the longest-lived Adélie penguins.[3]

The limitation on the availability of females of breeding age, resulting from their lower annual rate of survival at Cape Crozier, is consistent with trends in the yellow-eyed penguin and black-legged kittiwake.[3,62] That is, the sex having lowest survivorship begins breeding at an earlier age. The female begins breeding earlier in the two penguins, but the male does so in the kittiwake. In all three species, differences in adult survivorship between the two sexes relate inversely to the frequency and consistency of breeding effort, which is consistent with Wooller and Coulson's model,[62] as discussed earlier. It is highly possible, then, that the earlier onset of breeding in the less abundant sex (sex ratios start out equal) is a response to reduced competition for a superabundant resource (i.e., availability of the other sex). Conversely, increased competition for a limited resource (members of one sex) could select for deferred maturity in the other sex as a way to reduce the intensity of competition. Such a hypothesis has been implied in some studies.[31,57]

FOOD LIMITATIONS

By breeding in dense concentrations and thereby becoming central-place foragers (ch. 2), it is possible that seabirds experience a limited food supply within flight range of the colony because of intense intraspecific and interspecific competition.[9,23] Interference competition has been invoked to explain lower reproductive success in large seabird colonies compared with smaller ones of the same species.[24,27] In this situation, a threshold in foraging density is reached beyond which individuals begin to interfere with one another in the process of prey capture (e.g., two individuals dive for the same food item). It is possible that young, less experienced birds are the ones more easily excluded. Carrick[13] believed that in the royal penguin, older, established individuals could behaviorally exclude unestablished birds from food resources. A test of whether this actually happens among seabirds would entail measuring the age of first breeding as a function of colony size. It is not likely that such data are forthcoming, given the difficulty of acquiring such information even for one colony.

Food could also be limited by the penguins' ability to procure it. As discussed earlier, feeding efficiency must exceed that achieved during nonbreeding parts of the annual cycle for Adélie penguins to visit the colony and breed. Because young birds are less efficient feeders than older ones, they are prevented from breeding successfully.

ULTIMATE FACTORS

Following Lack,[30,31] Ashmole[9] believed that fecundity in seabirds is at its maximum (although with clutches of only one or two eggs, the norm for seabirds, it is certainly low) and that food availability near breeding islands was the major density-dependent factor regulating seabird populations. The low fecundity and deferred maturity necessitated more attempts at breeding and therefore a longer life span. Longer life certainly would be possible for most seabird species because breeding is not particularly hazardous for them (compared with the Adélie penguin) and because predation on adults is insignificant.

At this point in the discussion we have established some important facts: breeding is difficult for Adélie penguins, breeding even once increases the likelihood of breeding successfully in the next attempt, and reproductive maturity as well as breeding is deferred. To explain these facts we hypothesize that to make a breeding effort worthwhile, Adélie penguins must spend their first few years at sea learning to feed and then spend at least one year in the colony learning social skills. Because it is more difficult for males to enter the breeding population, more years as prebreeders are necessary. Lacking proficiency at feeding or social skills, Adélie penguins cannot breed successfully, so deferred maturity has arisen because individuals that have bred prematurely have not been successful. The cost of their efforts (often death) far surpassed the gains.

These hypotheses are much easier to accept with regard to the Adélie penguin than other seabird species because breeding for an Adélie penguin

Illustration 8.1 Adélie penguin chicks almost ready to fledge. *Drawing by Lucia deLeiris © 1988.*

is indeed hazardous. Therefore a premium exists on proficient breeding and feeding.[25] In the Adélie penguin, because of increased predation pressure, annual mortality of breeders, especially the youngest ones, is much higher than in nonbreeders.[3] Predation is also the major factor that selects against Adélie penguins breeding earlier than they do. By three years in females and four years in males, feeding is efficient enough that adequate physiological condition can be attained but only to such a degree that a few can breed successfully and avoid predation. Most Adélie penguins wait until they are even older. In the yellow-eyed penguin, which also lays and hatches two eggs, many females begin breeding at two years, but 68 percent of the eggs laid by two-year-olds are infertile.[46] If predation during the breeding season were a significant factor for that species and if young breeders were particularly vulnerable (as young Adélie penguins are), such wasted breeding effort certainly would be selected against. Based on pres-

ent data, it appears that Antarctic penguin species, particularly the Adélie, are unique among seabirds in clearly not being at the top of their respective food webs.

■ The Balance of Demographic Variables

Seabirds have some of the highest adult survival rates and the longest delays in attaining maturity among avian groups.[32,47,49] Deferred maturity is characteristic of all bird species with an adult annual survivorship of at least 60 percent, and as annual survivorship increases, so does the age at the onset of reproduction. A factor that is centrally important in these patterns is the very low reproductive rates of seabirds because most seabird species lay just one or two eggs during a given year.

Theoretically, some sort of compensatory changes should be possible to balance a population (births + immigrants vs. deaths + emigrants). This balance is far more evident in the evolution of a species' life history pattern than in the short-term perspective of a few generations of one species (i.e., the time span of the longest ecological studies). If mortality associated with reproduction is high (from an evolutionary perspective), it should be advantageous to reduce breeding effort and, in turn, fecundity (e.g., reduce clutch size or delay reproduction to ensure longer life).[59] If the breeding effort is reduced, the animal is better prepared to cope with factors causing mortality, and it is better to increase the number of breeding attempts than to rely on a few all-out efforts.[34,42] However, most seabirds are not subjected to predation as adults to a significant degree, and rarely has it been demonstrated that breeding is particularly risky for them. For example, reproduction is not considered to be especially hazardous for kittiwakes; nevertheless, the age of first breeding and the level and consistency of breeding effort among older birds balance annual production and survival rates.[62] Using the red-footed booby (*Sula sula*) as a model, Goodman[25] delved into this problem in great detail. He showed in theoretical terms how this seabird's low-intensity reproductive effort, which includes deferment of maturity, manifests the reproductive cost relative to the effort expended rather than restraint in reproductive effort as a means to achieve demographic balance, as other theoreticians have suggested.[46,47] Now that a number of long-term demographic studies of birds have been completed, the consensus seems to be that the life history strategy of a species is one in which demographic variables have been optimized (i.e., adjusted among themselves).[39,40,43] In other words, if adult mortality is high because of persistent environmental factors (e.g., predation), then fecundity must increase to compensate. Otherwise, the population or species would disappear. Whether increased fecundity takes the form of advances in the age of first breeding or increasing clutch size depends on other environmental factors. Also involved is the density of the population at certain life stages (e.g., a large number of chicks may not be possible given the food resources available to the parents).[40]

Information from four penguin species—Adélie, African, little blue, and yellow-eyed—all of which lay two eggs per clutch (fecundity equal), provides insights into how demographic variables compensate for one another in the larger evolutionary perspective. Especially evident is these species' inability to compensate imbalances on the time scale of a few generations (table 8.13). These are the penguin species on which much demographic information is now available; although it is not a given that the demography of all penguins should be alike, certain aspects of these species' natural history are similar. Most importantly, all are penguins and therefore do many things in a similar way for morphological and physiological reasons.[19] Three are similar in size (blue penguin is smaller), three are colonial (yellow-eyed is semicolonial), three are not very migratory (Adélie very much so), and three are inshore feeders (Adélie feeds far from shore). In studies of other seabirds, demographic variables have been found to vary according to body size (e.g., shearwaters[6] and, to a lesser ex-

Table 8.13 Demographic variables compared between four penguin species during periods when respective populations were increasing or decreasing.

Species	Location	Population Trend	Age First Breeding	Proportion Adults Breeding	Fecundity (chicks/pair)	Adult Survival
Yellow-eyed	Otago Pen, New Zealand	+	F: 2.5 M: 3.8	4 yr+, 0.81	1.15	0.87
Little blue	Phillip Island, Australia	−	F, M: 2.5	1.0	0.84	0.74
African	Robben Island, S Africa	−	F, M: 4.0	0.69–1.0	0.32–0.59	0.89
Adélie	Cape Crozier, Antarctica	−	F: 5 M: 6.2	8 yr+, 0.82	0.9	0.89

tent, penguins[19]). Thus the contrasts in demographic variables between these penguins, mostly of the same size, should be instructive. In three cases, the populations studied were decreasing (in two the decrease was compensated by immigration: little blue and African), and we consider them first.

In the little blue penguin,[20] adult survival was low, but age of first breeding was younger than in the other two decreasing species. The lower adult survival is consistent with the smaller body size but not to the degree shown; in this case, survival was affected by elevated predation on adults (from introduced mammals). In the African penguin, reproductive success was much lower than in the others.[17,49] Success was shown to vary according to the biomass of available forage fish (food availability), which in turn was affected by industrial fisheries and natural factors. Finally, in the Adélie penguin, the proportion of adults breeding was the factor that differed most, and as we have seen in previous chapters this difference is related to ice conditions. These sets of demographic parameters contrast markedly with those of the yellow-eyed penguin, for which the population was increasing at the time of study.[21,46] Proportion of adults breeding, breeding success, and adult survival were remarkably high, and age at first reproduction was young. Not surprisingly when we compare this set of values with those of the decreasing populations of the other penguin species, this penguin population increased 230 percent during the period of study.

Comparisons are rarely ideal, however. One should remember that youngest breeding yellow-eyed penguins lay mostly infertile eggs.[46] Therefore age at first breeding (as a mark of deferred maturity) in the yellow-eyed penguin is more similar to that in the Adélie penguin than it at first appears (that is, young birds that lay infertile eggs cannot really be considered mature). Richdale[46] thought that enhanced survivorship during winter and increased fecundity, both stemming from a density-related increase in food availability, accounted for the population increase. On one hand, the earlier age at first breeding might also be a response to increased availability of food. In other words, because of the initially low population density (from anthropogenic causes), food was more available (less competition between the penguins). On the other hand, early age of breeding and other factors could also relate to an absence of predation. Two pinniped species, which prey on yellow-eyed penguins,[52] may once have exerted greater predation pressure on the species than they did by the time Richdale conducted the study. Pinniped populations around

New Zealand (where the study took place) had been reduced tremendously by humans,[29] perhaps more so even than populations of the penguin. Thus age of first breeding may have responded to different pressures than annual survivorship, or the two variables may have responded independently to the same environmental factor. Because of protection, pinniped populations recently have been increasing around New Zealand. Does this mean that breeding by the yellow-eyed penguin may be more hazardous than in the recent past? It may be instructive to reinvestigate this species' demography. Different patterns may now be evident.

Other aspects of the Adélie penguin's life history patterns support independence of demographic variables but also indicate compensatory mechanisms. Adélie penguins take a few years to increase feeding efficiency to a point where they can visit the colony and additional years to increase efficiency to so that they can cope with predators and breed successfully. Thus food-finding ability (also a function of food availability and sea-ice extent) probably is important in limiting breeding potential by deferring maturity. Because predation reduces their life span, however, Adélie penguins must compensate by increasing clutch size and fecundity. Unlike other seabirds (including penguins) that also feed far at sea and take several days (one to three) to gather food and return to feed chicks, Adélie penguins lay and hatch two eggs instead of one and often raise two chicks to fledging. *Eudyptes* penguins, which also feed far at sea and therefore feed their chicks irregularly, lay two eggs but always lose one.[19] Thus a crested penguin never has the chance to raise more than one chick. Other penguin species that feed far at sea and make irregular visits to feed chicks (i.e., the emperor and king) lay only one egg. Therefore Adélie penguins exhibit a level of fecundity not typical of offshore-feeding penguins.

Any bird will produce the maximum number of young possible under given conditions.[30-32] Adélie penguins can support a clutch, and particularly a brood size, that is larger than that of other truly pelagic seabirds because of the tremendous increase in food availability during summer in Antarctic waters. The almost complete overlap in diet among Antarctic seabirds during summer is circumstantial evidence that food availability is not limiting during that season.[5,7] Further evidence for the ability of Adélie penguins at high latitudes to deal with increased brood size is their remarkably even production of chicks from year to year and from locality to locality, with rare indications of mass starvation. Such a catastrophe is common in tropical and temperate seabird species.[8,9,38] Additional evidence of high production is the age structure of an Adélie penguin population (at Cape Crozier), with concentration in the fledgling and early age classes. This pattern is a result of their high fecundity (two chicks raised).

The large number of prebreeding Adélie penguins relative to older individuals in the Cape Crozier population indicates that mortality must be heavy in younger age classes. This mortality probably results from starvation during winter, when populations of euphausiids and fish are minimal and when ice covering the sea surface reduces their availability even more.[60] Predation on Adélie penguins, particularly young ones, could also increase under such conditions because even leopard seals take advantage of the abundance of euphausiids during summer.[26] Under such conditions they ignore the chance to prey on penguins. With fewer krill available during winter,[36] however, the seals might then seek penguins as alternative prey. Mortality of young Adélie penguins during winter in relation to food availability thus could be an important density-dependent factor regulating population size. In fact, population growth is very sensitive to the survival of juveniles.[10,60] The same has been hypothesized for several other species of polar and subpolar oceanic birds.[31]

The actual mechanism by which food supply limits population growth in Antarctic plankton-feeding predators has not been considered extensively. The data on Adélie penguins in the Ross Sea presented herein are more complete than for Adélies elsewhere in the Antarctic. Extent of sea ice has much to do with food supply and is discussed in chapter 9.

Chapter 9

The Bellwether of Climate Change

The world's climate is changing, as it always has, but the rate of change is accelerating at an unprecedented rate, at least in the context of the last few thousands or tens of thousands of years.[24] More and more often, we read articles or hear stories in the news about global warming. Most climate researchers agree that Earth's surface air temperatures are warming at a rapid rate. Questions abound, and the answers reflect profound differences of opinion. Is the rate of change understandable as we continue to move away from the period of the Last Glacial Maximum (LGM)? If not, what explains the increasing rate of temperature increase? What changes in Earth's biota have occurred in response to the warming? Can we expect more of the same, or will rates of change in various species' populations accelerate or decelerate?

Antarctica has not escaped climate change. Well documented is the record for the last 420,000 years, which included four interglacial periods. In this record, the Holocene (which began 11,000 y.b.p., with the start of the present interglacial period) has to date exhibited more stable temperatures than the few immediately preceding interglacial periods.[38] Nevertheless, models of present-day climate predict that with increased levels of greenhouse gases, temperature increases will be greatest in polar regions.[34,39] Accordingly, over the last century surface air temperature, assessed in different ways from many localities but averaged for all of Antarctica, has risen 0.5–1.0°C depending on method of assessment.[11,28–30] There is no disagreement among researchers that a major increase has occurred. However, the trends vary by locality. For instance, little temperature change has been recorded in the interior of the continent.[11] Temperatures in the southernmost reaches of the Ross Sea have increased slightly, but they have increased 5°C or more at localities along the Antarctic Peninsula (fig. 9.1). Such an increase is a huge change. That is not to say that temperatures in Antarctica have become warm as we know them. Nowhere in Antarctica is the average annual air temperature above 0°C (not including remote islands in the northern periphery of the region). Nevertheless, the increases that have been recorded, especially in the northern part of Antarctica (i.e., the Antarctic Peninsula), are significant. Along with increasing air temperatures, ocean temperatures have warmed as well.[20] The temperature regime today along the west coast of the Antarctic Peninsula is unprecedented during at least the last 7,500 years (fig. 9.2).

A major consequence of increasing temperature for Antarctica has been disintegration of ice shelves in the Antarctic Peninsula region (fig. 9.3).[16,56] This is a major event. Not only has it been occurring for the first

Figure 9.1 A comparison of air temperatures between Faraday Station, Antarctic Peninsula (63°S), and Scott Base, Ross Island (78°S), 1957–1999. *Data from Smith et al.*[43] and New Zealand Antarctic Program.

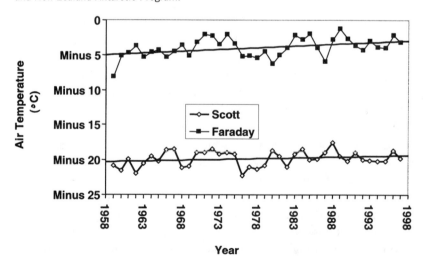

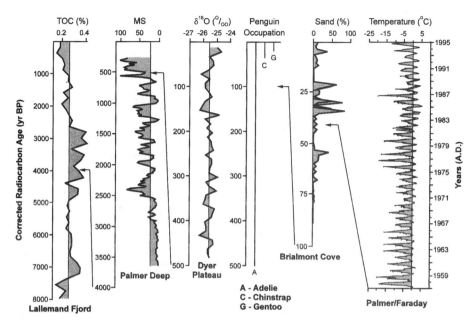

Figure 9.2 Evidence of climate change in the Antarctic, especially the west coast of the Antarctic Peninsula, as recorded in ice cores, coastal ocean sediment cores, and recent weather observations. TOC = total organic carbon in marine sediments; MS = marine sediment record showing magnetic susceptibility as a proxy for productivity. *Figure 3 in Smith et al.,*[43] *with permission © 1999 AIBS.*

time in recorded history, but disintegration has occurred in the space of one to two human generations; as much as 1,300 square kilometers of one shelf recently was lost in fifty days. Many of us have witnessed the disintegration first hand, which is very different from reading about the process as written in a geology textbook describing changes during the Pleistocene. When ice shelves (i.e., portions of continental glaciers that float on the sea) have retreated during the recent geologic past, slivers of adjacent coastal terrain have been laid bare, thus encouraging colonization by Adélie penguins. There are a few tiny colonies (i.e., those having fewer than 1,500 pairs) at the base of the Antarctic Peninsula, western side, that are in the vicinity of small ice shelves. These shelves have trouble existing where average annual air temperatures are warmer than −5°C.[56] These colonies, if dated, probably would be found to be very recent, a subject treated more fully below.

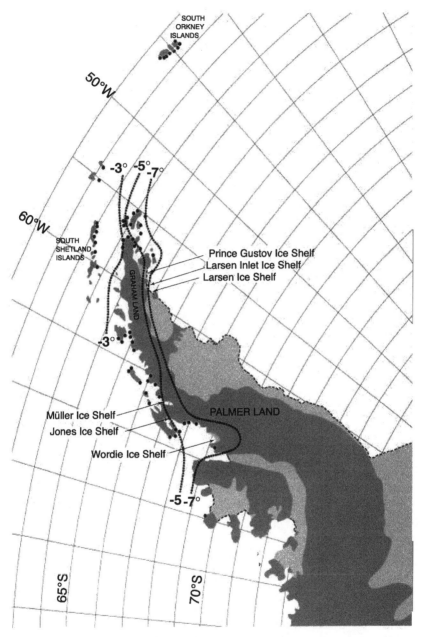

Figure 9.3 The Antarctic Peninsula showing location of present-day colonies in relation to ice shelves that have been retreating dramatically during the past forty years. *Colony data from sources identified in chapter 2; ice shelf data from Vaughan and Doake.*[56]

The retreat of ice shelves has not been observed continentwide. In the Ross Sea region, which is part of continental Antarctica much farther south than the Antarctic Peninsula, the eastern margin of the Ross Ice Shelf has been stable, but the western margin has been advancing.[31] The Ross Ice Shelf is one of the largest ice shelves on Earth.

Another consequence of recent climate warming, in accord with climate models of global warming,[26,39] has been a retreat in the amount and extent of sea ice. This has been most evident in the Bellingshausen Sea, which occurs along the western side of the Antarctic Peninsula[27,28] (fig. 9.4). Sea-ice cover also has been declining in the Amundsen Sea, immediately to the west of the Bellingshausen Sea. The decline correlates with the increasing air temperatures and may indicate larger-scale changes in wind and pressure systems in the southeast Pacific region. The decreasing sea ice in the ocean west of the Antarctic Peninsula is not representative of the remainder of the ice-covered portion of the Southern Ocean. Elsewhere, sea-ice extent has been stable or possibly increased, with much variability.[46]

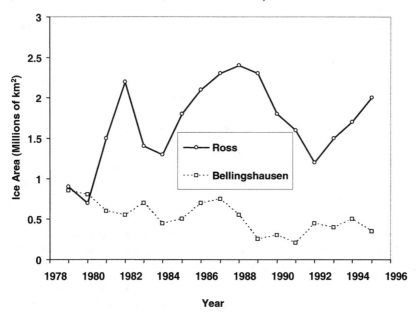

Figure 9.4 Changes in annual extent of sea ice in the Ross Sea and Bellingshausen Sea (west coast of Antarctic Peninsula). *Data from Stammerjohn and Smith.*[46]

Unfortunately, the sea-ice record is short. It begins only in 1973, when the first ice-sensing satellite went aloft. Instruments on these satellites sense the brightness of Earth's surface as transmitted by microwave radiation. The latter is short-wavelength radiation that reflects back to space, from the earth's surface, the long-wavelength radiation that reached it from the sun. Different textures of the earth's surface reflect radiation differently, and some substrates absorb more radiation or reflect more of it back than do others. These differences are evident in the satellite images prepared from the data. Using data collected at the earth's surface (e.g., from a ship) as a satellite passes overhead, statistical formulas have been developed to translate the satellite's readings into ice cover or the amount of open water. This is called ground-truthing of the brightness readings that the satellites provide. In other words, an ocean totally covered by ice would be very bright, and open water would appear dim. Most recent assessments of ice cover include only data collected since 1978, when a multisensor satellite was successfully launched. Before then the satellite had only one brightness sensor. The added sensors have allowed cross-correlations to be made as a way to calibrate what the readings represent. In any case, twenty years is not very long as climate records go.

A longer-term record of sea-ice extent during this century—if we want to consider indirect measures—is provided by records of whaling ships.[15] Researchers often substitute one measurement that is easy to make for another that is difficult or impossible to make as long as there is a strong correlation between the two variables. After breeding in warmer surroundings, baleen whales migrate to the Southern Ocean during austral spring and summer to feed on the abundant krill. The whales remain for several months, feasting all the while.[9] The whales move southward as the sea ice retreats, exposing more krill to predation.[33] In turn, the whaling ships have followed the whales, although no farther than the pack-ice edge. As whales were killed, the positions of whaling ships, particularly factory ships, has been entered into trip logs. Analysis of the positions, dating from the 1930s, shows little change in where the whaling area occurred through the early 1950s. Then there was an abrupt increase in latitude (southward movement of the whaling area) until the early 1970s.[15] Subsequently, positions of whaling ships did not move farther south, indicating stability in the position of the pack-ice edge from the early 1970s on. Unfortunately, the satellite record begins only in 1973, as noted earlier. Thus there is little overlap between satellite data and whaling ship data. However, satellite data also show stability in ice extent since 1973 (except in the waters west of the Antarctic Peninsula).

There appear to be some problems with the whaling ship analysis.[1,57] For one thing, species-specific whale data were combined regardless of whale species. In the latter part of the whaling era, the exploitation switched from the great whales (because few remained) to the smallest baleen species, the minke whale. The latter species is far more common in the sea ice zone than the other baleen species (see ch. 2, discussing this whale's relationship to Adélie penguins).[9,32] Therefore the apparent southward shift in whaling position could be related more to this shift in target species. Another problem with the whaling ship[15] analysis is that the supposed 2.8° southward shift in the outer ice edge is just not believable given present-day variation of the ice edge position and variation during the LGM.[10] It may be that the shift actually measured was the rate of seasonal decrease from maximum.[1] A much faster ice retreat during modern times would bring the ships farther south sooner in the summer than earlier in the century and would account for the apparent difference. Regardless, the long-term Adélie penguin population trends in the Ross Sea, complementing those of emperor penguins from Adélie Land,[4] indicate that something did change in the qualities of sea ice between the middle and later part of the twentieth century. What the penguin records indicate about changing ice conditions is discussed more fully later in this chapter.

Speaking of baleen whales, their loss from the Southern Ocean marine ecosystem was huge, numbering in the hundreds of thousands. The baleen whale population (several species) currently is about 35 percent by number and 16 percent by biomass of its preexploitation abundance.[8,32] These percentages by number and mass differ so much because the whalers took the larger whale species (with greater biomass) preferentially. Baleen whales feed almost entirely on krill in the Antarctic, a main food item in the diet of Adélie penguins that frequent continental slope habitats. Therefore the loss of so many whales constitutes a major change in the marine ecosystem during this century, a change that could well have had repercussions for Adélie penguins. After all, the whales could have competed with penguins for food.

■ Trends in Adélie Penguin Populations

In the context of environmental change, have Adélie penguin populations exhibited corresponding trends? If so, has there been a consistent pattern in trends of Adélie penguin populations in different regions? The answer to the first question is yes; that to the second is no, and in ways that at first glance appear to be contradictory. Can we explain the observed changes in the populations? This question is addressed later in this chapter.

Numbers of breeding pairs of Adélie penguins have been determined at Cape Royds, Ross Island, annually since 1959.[50–52,58] This is by far the longest continuous record in existence to show annual variation and trends in a single Adélie penguin colony. It is one of the longest biological records anywhere in the Southern Ocean. Elsewhere on Ross Island, the size of the colony at Cape Bird has been assessed annually since the mid-1970s, and the one at Cape Crozier since 1983; counts were done at both colonies infrequently in earlier years.[51] Therefore we know fairly well the recent population history of most colonies in an entire metapopulation (a demographically related group of colonies isolated from others). In this cluster of colonies, Cape Crozier is the largest (120,000

Figure 9.5 Changes in numbers of breeding pairs at three Adélie penguin colonies on Ross Island, 1958–1998. *Data from Wilson et al.*[58]

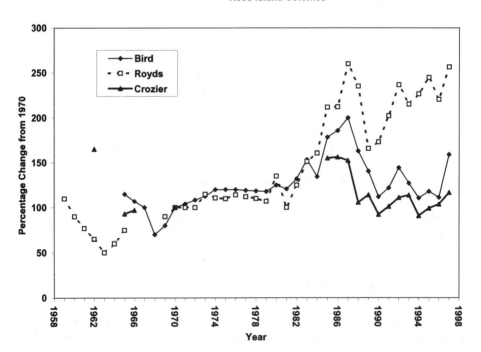

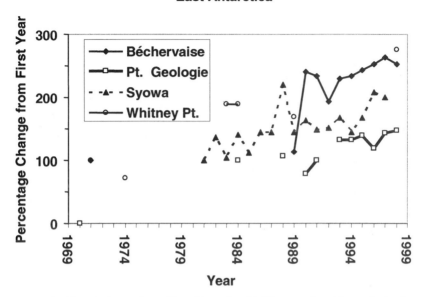

Figure 9.6 Changes in numbers of breeding pairs of Adélie penguins at four colonies located along the Indian Ocean sector of Antarctica, assessed fairly regularly since about 1970. *Permission to present these data as follows: Béchervaise Island, J. Clarke; Syowa area, A. Kato; Point Géologie, H. Weimerskirtch; Whitney Point, E. Woehler (some data reported elsewhere).*[36,59,60]

pairs), Cape Royds the smallest (4,000 pairs), and Cape Bird is midway in size (36,000 pairs; table 3.1).

Elsewhere, information on changes in colony size is available for several colonies, equivalent to the record from Cape Crozier.[14,21,35,53,60,61] With only an occasional estimate of size during earlier years, the time span of the mostly annual data for these colonies is as follows (table 3.1, fig. 3.1): Indian Ocean coast, Syowa Coast (several closely spaced colonies) 1981–1997; Bellingshausen Sea coast, Arthur Harbor, Anvers Island 1978–1995; Weddell Sea, Signy Island, South Orkney Islands 1979–1998[60] (figs. 9.5, 9.6). Other useful data come from Béchervaise Island, Point Géologie, and Whitney Point, all on the coast of the Indian Ocean sector (East Antarctica; fig. 9.7).

In the southern Ross Sea, Adélie penguin colonies decreased in size from the late 1950s until about 1970, then began a slow increase (fig.

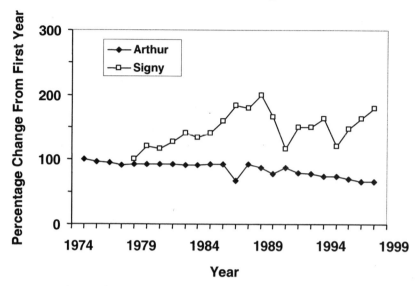

Figure 9.7 Changes in numbers of breeding pairs of Adélie penguin colonies located in the Antarctic Peninsula region, assessed fairly regularly since about 1980. *Permission to present these data as follows: Arthur Harbor, W. Fraser; Signy Island, A. Lynnes, J. Croxall.*[60]

9.5). The increase accelerated in about 1982, then reached a plateau in the late 1980s. A crash then occurred in 1989. Since then, the Cape Royds colony has recovered the 1987–88 peak, Cape Bird has almost recovered, and Cape Crozier has not. Since 1980 or so, when their respective records began, colonies on the Syowa Coast and those at Signy Island have shown a roughly similar pattern to that at capes Royds and Bird (cf. figs. 9.5–9.7): growth to the late 1980s, a crash, and then slow recovery but still at a higher level than when the census record began (ca. 1980). The data from other colonies of the Indian Ocean sector are missing the critical years in the late 1980s (with the crash and recovery), but otherwise the trends are consistent with Ross Island, Syowa, and Signy Island.

In stark contrast are the histories of Adélie penguin colonies located off the west coast of the northern Antarctic Peninsula (fig. 9.7). At Arthur Harbor, numbers of Adélie penguins have declined since the late 1970s, especially after 1990. These declines are consistent with a number of abandoned Adélie penguin breeding sites located in the vicinity of Arthur Harbor.[18]

Factors Explaining Population Trends

RETREAT OF GLACIERS

Major ice shelves, which are floating extensions of land-grounded continental glaciers, have disappeared at several locations on the west coast of the southern Antarctic Peninsula and near its northeast tip since the 1940s (fig. 9.3).[56] The retreat begins when, in the vicinity of an ice shelf, the mean air temperature during January reaches 0°C or the annual air temperature rises above −5°C. To affect Adélie penguins, the retreat of an ice shelf must be large enough that some coastal terrain is laid bare as well. This would not necessarily occur immediately. Although the colonization of sites where ice shelves have retreated in the Antarctic Peninsula region has not been documented, this phenomenon is well documented in the recent geologic record elsewhere in the Antarctic. Soon after the land was exposed, the penguins seemed to have found it. The term *soon* is defined in geologic time, and whether that means a century (or more or less) is not known. Maybe in the near future, given the rapid retreat of ice shelves in the Antarctic Peninsula region, we will have the chance to find out. Some sort of glacial rubble or till (small stones) would have to have been left behind for the penguins to use in building nests. If the glacier scrapes the basement rock free of a gravel layer, the penguins cannot establish a colony.

As far as is known, no new Adélie penguin colonies have been founded as a direct result of the well-documented recent retreats of glacial ice in the Antarctic Peninsula region. The areas involved—Bellingshausen Sea coast south of 70°S and Weddell Sea coast south of 64°S—have very few Adélie penguin colonies to begin with (figs. 3.1E, 9.3). Adélie penguins probably have not founded new colonies because of the compacted sea ice that exists in these areas year round and the corresponding lack of significant polynyas. Thus access to the sea from prospective colonies would be a problem. Given the record of temperature increase over the twentieth century in the Antarctic Peninsula region, however, good candidates for recently established colonies as a result of glacial retreat are (table 3.1; figs. 3.1, 9.3) as follows: Rhyolite Island (35 pairs; colony 92, table 2.1), near the northern front of the retreating George VI Ice Shelf; Lagotellerie

(1,700 pairs; colony 94); Pourquoi Pas (700 pairs; colony 95) islands, near the outer front of the retreating Jones Ice Shelf; and Andressen (2,200 pairs; colony 101) and Detaille (900 pairs; colony 101) islands, near the outer edge of the retreating Müller Ice Shelf. It is highly likely that these colonies are very young.

DISSIPATION OF SEA ICE

To understand the possible role of sea ice in the population trends exhibited by Adélie penguins in recent years one needs to understand the relationship between ice extent or concentration and this species's breeding and foraging ecology. It has been pointed out many times in this book that Adélie penguins are obligate associates of sea ice, and much attention has been given in earlier chapters to how variation in sea-ice conditions affects this species. In other words, this species is not found either at sea or at colonies where sea ice is not an annual feature of the environment. This species also is not found where sea-ice concentration is at the other end of the spectrum: extensive fast ice or compacted sea ice present through much of the year. About 15 percent at the low end and about 85 percent at the high end are the approximate limits of suitable ice cover, although much further study is needed to better define this effective range. The relationship can be illustrated by a conceptual model (fig. 9.8).[43]

The reasons why Adélie penguins do not do well when fast ice or compacted sea ice is present are well understood. This subject has been treated exhaustively in the discussions of breeding biology in chapters 4–6. In summary, fast ice (which constitutes 100 percent ice cover) or pack-ice cover greater than about 85 percent forces Adélie penguins to walk over the ice. As discussed in chapter 2, they walk or toboggan at a speed of 1–2 kilometers per hour, much slower than their rate of travel when in long-distance swimming mode, 7–8 kilometers per hour. This means that they cannot commute between nest and feeding area often enough to effectively satisfy the needs of mate or chicks (mates get hungry and desert the nest, chicks starve). It also means that they consume food for self-maintenance that otherwise could be given to chicks (thus chick development is retarded). With 100 percent ice cover, of course, penguins cannot hold their breath long enough to forage very far under the ice. Prey availability becomes depleted quickly within a breath-hold distance of the ice edge. Ice cover at 100 percent precludes the existence of any air-breathing vertebrate (except Weddell Seals, which gnaw at the ice to keep breathing holes in tide cracks open).

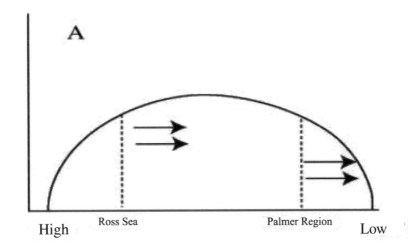

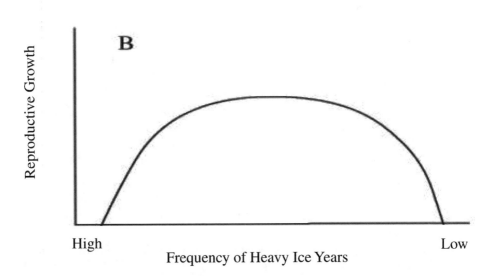

Figure 9.8 A) The conceptual model presented by Smith et al.[43] showing the relationship between Adélie penguin well-being (a measure of reproductive effort or success) and sea ice concentration. B) A modification of the Smith et al. model, applicable to population growth and size.

In a recent analysis of the population trends of Adélie penguin colonies in the Ross Sea, growth was inversely proportional to sea-ice extent during winter with a five-year time lag.[58] In other words, extensive sea-ice cover during July–October (time of maximal ice extent each winter) caused a decrease in the size of breeding populations several years later. Conversely, a winter of lower-than-average ice extent translated to a population increase several years later. The explanation of this relationship hinged on overwinter survival of juveniles and subadults, which take several years to recruit into the breeding population. A hard winter severely depleted the numbers of young, potential recruits, much more than adults, which probably have already learned to cope. Overwinter survival was discussed more fully in chapter 8.

It is thought that the greater ice extent moves the outer portion of the pack ice beyond the waters lying just south of the Antarctic Circumpolar Current (ACC). As noted in chapter 2, these waters are very rich in plankton and therefore penguin food.[37,55] The waters north of the southern boundary are much less rich. Finding food there would be more difficult, and the inexperienced juvenile penguins would fare the poorest. In addition, predation of penguins by leopard seals might also increase if there were insufficient alternative prey for them (i.e., krill). The leopard seals, being air-breathers themselves, also dwell in the outer, looser portion of the pack ice during winter. Greater susceptibility to predation of young penguins was discussed in chapters 7 and 8.

At the other end of the spectrum, it is not well understood why Adélie penguins do not do well if there is no sea ice. This subject was treated in chapters 4–6. Adélie penguins are much more capable of exploiting the opportunities presented by ice-covered seas than their relatives. One useful attribute is that they can hold their breath longer than other species of similar size. This ability allows them to search for organisms longer on the underside of ice floes. It is possible that Adélie penguins could be at a competitive disadvantage wherever sea ice is absent. Such conditions characterize the habitat of the world's most abundant penguin species: macaroni, chinstrap, and rockhopper penguins. In simple terms, for want of anything better, an Adélie penguin is associated with pack-ice-covered seas, just as a duck is found where there is water. Where there is no ice, there are no Adélie penguins.

Among the sectors of Antarctica where sea ice is a year-round feature, namely those in the Indian Ocean sector, the Ross Sea, and the northwestern Weddell Sea, the pattern of colony growth since the late 1970s has been similar, at least in the portions of records that overlap in time. All populations grew until the late 1980s. Then they exhibited a slight crash over a one- to three-year period, followed by much variation as they began to recover slowly. The smallest colonies—those at Cape Royds (4,000 pairs) and Syowa Coast (fewer than 2,000 pairs)—showed the most resilience to the population crash. In both cases, the most recent estimates of population size equal those attained in the late 1980s. Thus recovery has been completed, and growth probably is continuing. Other, midsized colonies are moving toward recovery but have not yet recovered (as of 2001). Very large colonies (e.g., that at Cape Crozier) have shown only slight growth since the crash. Therefore, it appears that the patterns exhibited in the long record at Ross Island[58] apply widely in continental Antarctica.

It also appears that the dynamics of adjacent colonies are involved in growth of individual populations, assuming that the Ross Island colonies can be used as a model. The small Cape Royds colony (full recovery from the crash) and the medium-sized Cape Bird colony (partial recovery) are satellites of the large Cape Crozier colony. It appears that after 1990, conditions favored the peripheral colonies over the large source colony. In fact, after 1990, there were several years in the southern Ross Sea in which sea ice was minimal during the nesting season. It was found that under such conditions, the penguin parents at Cape Crozier had to work much more to rear chicks than those at Cape Royds because sea ice (and thus the feeding area) was much farther away for the Crozier birds. Crozier chicks fledged at a much lighter mass, and the parents lost condition.[3]

The analysis relating Adélie penguin population size on Ross Island to ice extent in winter[58] covered only the period 1979–1998. Even though penguin population size at Ross Island colonies was known earlier than 1973, information on sea-ice extent was unavailable earlier because the first sea-ice sensing satellites were launched in 1973. That is unfortunate. However, it is possible to make some useful conjectures about what was happening between 1957 and 1972. During this time, the colonies at Cape Royds and at Cape Crozier were declining[2,52,58] (fig. 9.5; see also ch. 8). Assuming that the analysis by Peter Wilson and colleagues[58] and correlations to sea ice apply, winter sea ice must have been more extensive or concentrated during this earlier period. In a way, this pattern is corroborated by emperor penguins nesting in Adélie Land.[4] Like that of the Adélie penguins of Victoria Land, the emperor penguin population was stable during the 1950s and 1960s, but in the 1970s, it began to decrease. Since then, the population has stabilized but at a lower (more variable) level. The emperor penguins nest during the winter and benefit from reliably heavy ice conditions (in fact fast ice), the opposite condition that Adélie penguins find to their liking. More work is needed to further relate the trends shown by these two species.

Previously, it had been thought that overly frequent visits by tourists to the Cape Royds colony led to the population decrease during the 1950s and 1960s.[52] The colony is very close to McMurdo and Scott bases and therefore became a regular destination for people wanting to complete their "Antarctic experience" by seeing penguins. In retrospect, however, we now know that other colonies in the region were decreasing at that time, too (fig. 9.5). Tourists did not visit these other colonies much. Thus the tourist visits, if anything, were contributing to a pattern already under way but caused by other factors, as in the more recent experience at Arthur Harbor. There, the part of the colony visited by tourists was stable in size, but the part designated a Specially Protected Area (no tourists and very limited visits for scientific purposes) has all but disappeared.[21]

Among colonies located along the western side of the Antarctic Peninsula, the pattern of colony growth has been consistent as well. In this case, though, unlike the colonies located where sea ice is a constant habitat feature, the pattern has been one of gradual decline, with decline becoming more steep after 1990. In this case, the lack of sea ice is the limiting factor (fig. 9.8). A number of studies have indicated a dramatic reduction in ice cover in the waters west of the Antarctic Peninsula during the past two decades. It appears that without sea ice, Adélie penguins are finding this area unsuitable.[18] In turn, other penguin species, other seabirds, and marine mammals that are not associated with sea ice are invading this area from the north.[22,43] In years of heavier-than-normal winter sea-ice cover, at least for this region, Adélie penguins have exhibited their highest reproductive success.[53,54] However, extensive ice cover for this region is sparse ice cover for the Ross Sea and many other parts of Antarctica.

FOOD AVAILABILITY

As pointed out earlier, the demise of whales in the Southern Ocean ecosystem should mean that more krill are available to other species that also consume krill. If so, one potential result should be a coincident increase in numbers of krill predators other than whales. The populations of certain species, specifically the chinstrap penguin, have increased dramatically since commercial whaling was curtailed. It is conceivable that the disappearance of whales is a factor,[33] as is the dissipation of sea ice.[13,14,22] In the case of the Adélie penguin, which resides in a habitat avoided by the great whales (pack ice), it is unlikely that the disappearance of whales had much effect on the penguin's population trends. The whales, like the open-ocean penguin species, tended to avoid the sea ice, where the Adélie penguin feels at home.

■ Patterns Revealed in the Prehistoric Record

The genus *Pygoscelis* appears in the fossil record at the end of the Pliocene, or about 3 million y.b.p.[41] Thus pygoscelid penguins are no older than humans. As far as is known, the first pygoscelid was very similar to the gentoo penguin and lived in New Zealand. This suggests that the climate of the Southern Ocean was colder then than now because gentoo penguins no longer exist that far north. The Pleistocene, which followed the Pliocene and lasted to about 11,000 y.b.p., was marked by a series of glacial periods (four major ones) both in the Northern and Southern hemispheres.[38] We now live in the Holocene, which covers the interglacial period (presumably) that followed the last Pleistocene glacial advance (Wisconsin period). Thus the pygoscelid penguins have been dealing with alternating periods of cooling and warming throughout their existence. How these penguins, particularly the Adélie penguin, have been responding to this variation helps us understand present-day changes.

We know something about how Adélie penguins responded to glacial events—the founding and abandonment of colonies—because the age of subfossil remains has been determined using a method widely used in geology called carbon dating. In nature, carbon-14 (^{14}C), a naturally occurring isotope, is incorporated into the tissues of all living creatures. This isotope is formed when CO_2 in the atmosphere is bombarded by cosmic rays (which is happening all the time). The CO_2, with ^{14}C as the carbon atom, is incorporated into plants (marine algae) through photosynthesis. The amount of ^{14}C in living tissue is known, and at a known rate it is magnified up the food chain as one animal eats a plant, the next animal eats the first animal, and so on. Once the animal dies, ^{14}C no longer is added but degrades at a known rate into stable forms of carbon. Thus the amount of ^{14}C in a bone or egg fragment, compared with what the amount would be in a living animal, indicates how many years it has been since the animal died. The result is approximate (say, within a hundred years or so for a Holocene bone) for a number of reasons. Confounding factors include contamination of samples by an atmosphere enriched in ^{14}C by nuclear tests (which generate ^{14}C) conducted since the 1940s. Thus ages appear to be younger than they are. Another factor to contend with when considering the age of marine organisms is the incorporation of "old" carbon that was deposited in the calcium carbonate remains of microscopic organisms a very long time ago but only recently has been dissolved and upwelled to the surface in deep water. This old carbon is depleted in ^{14}C and thus makes organisms that incorporate it seem older than they are. Geologists have worked to establish correction factors to account for erroneous carbon dating, but the science is still uncertain. At present, all that geologists can do in this regard is to agree on what correction is the best.

In terms of temperature fluctuation in the Antarctic, the Holocene has been much more stable than other recent interglacial periods.[38] In response to the warmer temperatures and retreat of ice sheets, the Adélie penguin recolonized much of the continent's shores. The history of this process is best known for the Ross Sea. There, and elsewhere to a much lesser extent, geologists have dated bones (and other penguin-derived material) found in the soil beneath colonies to estimate when the penguins first nested there. This provides an estimate of when the ice sheet retreated to provide nesting habitat. Of course, the geologists were interested in the ice-sheet chronology rather than the penguin chronology. Curiously, to date no subfossil bone has been found (dated) from an Antarctic penguin that lived before the LGM (i.e., the Wisconsin Glaciation). In other words, we have the fossil pygoscelid bones that have been dated to 3 million y.b.p., but the next oldest pygoscelid bones are subfossil and are dated to about 12,000 y.b.p. Why nothing in between? Possibly, older bones have been scraped from presently exposed coastal terrain by advancing glaciers and ice caps during the intervening years and have been ground to dust or lie on the sea floor. This is one explanation offered to account for the lack of bones older than 700 y.b.p. in the Antarctic Peninsula region.[18,49] Or (and this is a more likely possibility) the appropriate terrain has yet to be searched. However, such terrain is extremely limited.

In the pebbles and sand that compose beaches currently exposed around the coast of Antarctica, there is an abundance of marine subfossils and fossils (but no penguin material) originating 35,000–20,000 y.b.p.[7] This is an interglacial period when—as we know from nonpenguin sources of information—coastal glaciers and ice caps were in a phase of major retreat. Given that coastal areas were ice free, Adélie penguins should have been nesting along the Antarctic coast during this period. During the period 20,000–8,000 y.b.p., coinciding with the LGM, little of such marine material (e.g., clams, algae) was deposited on coastal beaches, but it has been deposited since. This indicates that during the LGM, most of the coast of Antarctica was covered by ice. It is likely that few, if any, Adélie penguins used colonies located on the Antarctic mainland during this time, given their need for pebble-rich, ice-free terrain.

Where the Adélie penguins nested during the LGM is not known, but presumably they did so on islands well north of the current distribution of this species. During the LGM, sea ice occurred several degrees farther north in the Southern Ocean than it does today; the Polar Front also occurred farther north than it does now.[10] This is known from the amount and species composition of diatoms whose calcareous skeletons have sunk to the bottom of the sea, as mentioned earlier. They are found as layers in deep-sea cores, the diatom fauna changing with the extent and persistence of sea ice overhead. The results of deep-sea core analysis indirectly indicate that certain islands that now occur north of the sea-ice zone (and where Adélie penguins do not now nest) could have been within the sea-ice zone during the LGM. Thus Adélie penguins could well have nested at these sites at that time. Candidate sites, which today are close to the northern extent of sea ice, might include South Georgia and Heard Island. These islands, almost sub-Antarctic today, are covered in vegetation, so any abandoned nesting sites, if not covered by colonies of present-day penguin species, probably are obscured by carpets of mosses and grasses. Therefore no prehistoric, abandoned nesting sites have been found from which to obtain and date any subfossil bones that might be present.

Adélie penguins could have nested among the South Sandwich Islands during the LGM. They nest in large numbers among those islands today, and the sea ice is present near them probably for the barest minimum portion of the year needed to suit this species. It is also possible that Adélie penguins nested on the continent at Cape Adare, Victoria Land, which may have been exposed during the LGM. At the least, the grounding line of the West Antarctic Ice Sheet (WAIS) was sufficiently far south that Cape Adare should have been free of the ice sheet, in contrast to almost all the remainder of Victoria Land (and the continent).[6,48] Adélie penguins also could have nested at the South Shetland Islands, which were not reached by the WAIS either. However, these islands, like the South Orkneys, may have had their own ice cap, so ice-free coastal land may not have existed.[25]

COLONY FORMATION AND ABANDONMENT

Studies using carbon dating of penguin remains in the Ross Sea region indicate that Adélie penguins colonized sites soon after they were exposed by the retreating ice sheet after the LGM (28,000 to 19,000 y.b.p.).[12] The WAIS (in effect, the Ross Ice Shelf) receded southward and eastward—behaving like an inward-swinging gate hinged in the eastern Ross Sea (Marie Byrd Land)—to reach its present position near Ross Island, southern Ross Sea, approximately 7,750 ± 90 y.b.p.[6,48] The grounding line was north of Ross Island as late as about 9,400 y.b.p.[12] The oldest penguin remains in the southern Ross Sea, among the colonies dated, have been estimated at 7,070–8,080 ± 180 y.b.p. (fig. 9.9).[6,44] These remains came from Cape Bird, the northernmost point of Ross Island (table 3.1, fig. 3.1). In Terra Nova Bay, Victoria Land, 300 kilometers northwest of Ross Island, habitat became available for nesting at about the same time,[48] as confirmed by carbon dates: about 7,065 ± 250 y.b.p.[6] (colonies 73–74 in table 3.1). Adélie penguin remains at Cape Royds, Ross Island, near the present position of the ice sheet front but farther south than Cape Bird, were dated at 442 ± 75 y.b.p.[5] At Cape Barne, the currently ephemeral colony a little farther south, uncorrected dates indicate colonization at 375 y.b.p.[45,47] At Beaufort Island, just 20 kilometers north of Cape Bird, Ross Island, the oldest known remains are young: 1,150 ± 45 y.b.p.[5]

When the ice sheets reached maximum size, their mass pushed the Antarctic land mass downward. With warming or whatever processes contributed to retreat of the glaciers (e.g., rising sea level facilitated retreat of ice shelves), as the sheets decreased and lost mass the continent emerged. Many Adélie penguin colonies now exist, including many reviewed earlier, on the raised beaches, now well above sea level, that remain from successive periods of ice retreat. In particular, researchers at Terra Nova Bay[5,6] dated penguin bones to determine the glacial and geologic chronology of the area's formation and emergence.

Along the coast south of Terra Nova Bay, Adélie penguin remains have been found thus far at six sites where the species does not nest today[6] (fig. 9.10). Nest stones also were found, indicating that the species actually bred at these sites. Five of these sites were occupied 2,900–4,300 y.b.p. As

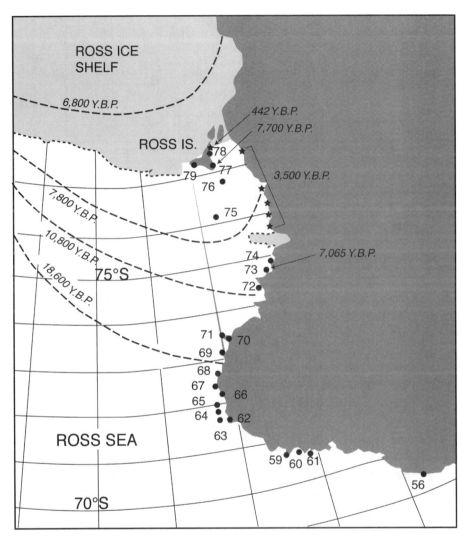

Figure 9.9 The Victoria Land coast showing oldest dates of currently active and abandoned penguin colonies. Dashed lines represent the grounding line (and approximate ice front) of the West Antarctic Ice Sheet (in effect, the Ross Ice Shelf).[12] Stars indicate abandoned colonies; dots indicate active colonies, with numbers corresponding to those in table 3.1 and figure 3.1. *Adapted from Baroni and Orombelli[6] and other sources identified in the text.*

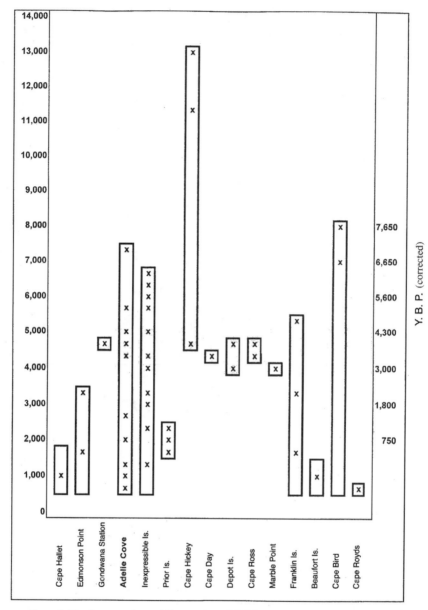

Figure 9.10 A comparison of the number of Adélie penguin colonies that existed along the southern Victoria Land coast during the Holocene, based on how the term *rookery* (or *colony*) is defined. The "colony" at Gondwana Station may actually be a part of the Adélie Cove colony as well; it is barely beyond the distance (8 kilometers) used to define colony extent. *Data from Baroni and Orombelli.*[5,6]

discussed in chapter 3, this coast is no longer suitable for Adélie penguin nesting because of the persistence of a several-kilometer-wide strip of fast ice that breaks out only every few years. The exception is Terra Nova Bay itself, which is blasted by offshore, katabatic winds that limit the presence of persistent ice and produce a small polynya. On the basis of penguin remains, this polynya has been present for at least the last 7,000 years. Because of the existence of the several now-abandoned colonies along this coast, the period 2,900–4,300 y.b.p. is now known as the penguin optimum.[6] Either air (and sea-surface) temperatures were warmer then or winds were stronger. In fact, this was a warm period, as identified in sediment cores collected to the east along the Bellingshausen Sea coast (fig. 9.2). At about 2,600 y.b.p., at least according to the sediment records, Antarctic coastal temperatures cooled. This date corresponds to the age of the youngest penguin bones in the abandoned Victoria Land sites.

Unusually old penguin remains (guano) were found at an abandoned site on Cape Hickey, about 100 kilometers south of Terra Nova Bay. The dates, 11,325 ± 360 y.b.p. and 13,070 ± 405 y.b.p., are two to three times older than any other Adélie penguin material in the southern Ross Sea or elsewhere.[6] The WAIS grounding line, estimated from data sources other than the dating of penguin remains, occurred at Terra Nova Bay at about 10,825 ± 640 y.b.p. Therefore something is amiss or poorly understood about the glacial retreat and uplift of the region or, more likely, the dating of the Cape Hickey deposits.[12] If Adélie penguins occupied Cape Hickey then, why did they not occupy other sites just a short way north and south? These other sites were colonized first at about 7,000 y.b.p. To pose the question in a different way, why has material of the older age not been found at other sites?

Something is indeed amiss: Dating penguin remains, particularly if sampling was less than exhaustive, provides only minimal dates of occupation. Dates of material from Cape Hallett, at 1,210 ± 70 y.b.p.,[6] and at Cape Adare, at 650 ± 90 y.b.p.,[5] almost certainly are gross underestimates of when penguins first established these colonies. Capes Adare and Hallett both were north of the WAIS grounding line during the LGM, and Cape Adare may have been ice-free since 35,000 y.b.p. Part of the ice sheet during the LGM could have floated beyond the grounding line as an ice shelf to obscure land at Cape Hallett. This is less likely for Cape Adare, where diatoms in the ocean sediments offshore indicate that sea ice, as opposed to shelf ice, was present at the sea surface at that time.[48] On the other hand, maybe these sites really were only recently occupied by Adélie penguins because other, unknown factors rendered these capes unsuitable for penguins during the LGM.

Beaches also have been dated to estimate timing of glacial events in a study of the Windmill Islands, Budd Coast (colonies 39–42, table 3.1). These islands, in the Indian Ocean sector of Antarctica, were completed covered by the East Antarctic Ice Sheet (EAIS) during the LGM.[23] Glaciers disappeared from the southern islands about 8,000 y.b.p. and the northern ones about 5,500 y.b.p. Some penguin material from a site in the northern area was dated to 4,280–4,530 y.b.p. This is a minimum age, and it was thought that the colony probably was 500–1,000 years older, which would correspond to when the nearby raised beach was exposed (the penguin material comes from what was a submerged pool at that time). Therefore, as in Victoria Land, soon after the sea and land were free from glaciers, Adélie penguins established colonies. Actually, it is possible that the only ice-free section of coast in East Antarctica during the LGM was at Lützow-Holm Bay.[25] Today, the Adélie penguin colonies there are very small because of the presence of compacted sea ice in the region (colonies 2–8, table 3.1). There is no reason to believe that the sea-ice regime has changed there, but we don't really know.

Adélie penguin remains from the Antarctic Peninsula region tell an even more complex story than that of continental Antarctica. On the basis of age estimates derived from sediments and other means, the following is a sampling of the periods when certain areas, most now occupied by Adélie penguins, became glacier free (minimum estimates): Bransfield Strait, 14,000 y.b.p.; Elephant Island, 5,500 y.b.p.; King George Island, 8,400–5,500 y.b.p.; Livingston Island, 5,000–3,000 y.b.p.; James Ross Island and islands nearby to the north, 7,300–4,500 y.b.p.; Anvers Island, 6,000 y.b.p.; Adelaide Island, 8,000 y.b.p.; Alexander Island, 6,000 y.b.p.[25] There is no evidence that the South Shetland Islands were overridden by continental ice, but possibly they were covered by ice caps (as noted earlier). The oldest mollusks found on raised beaches of the South Shetland Islands were deposited 8,400 y.b.p. Such beaches should have been available for nesting by Adélie penguins. Some readvance of glaciers occurred at various times since the LGM, but how this may have affected Adélie penguins is not known.

Consistent with these chronologies of glacier retreat are the dates of penguin remains from Lagoon and Ginger islands, Marguerite Bay, at the base of the Antarctic Peninsula (colony 99 in table 3.1; 67°45′S, 68°42′W).[19] These sites are in the vicinity of Adelaide Island and just south of Anvers Island. The oldest remains at Lagoon Island were dated 6,000 y.b.p., with intermittent occupation thereafter. These results support the pattern, observed elsewhere, that occupation by Adélie penguins occurred soon after the retreat of the ice sheets that were in place during the LGM.

It is possible that readvanced glaciers have destroyed many of the older penguin remains in the Antarctic Peninsula region or, less likely, that key sites there are obscured by the lush growths of mosses and grasses present today in many places.[18,49] Indeed, except for Marguerite Bay, it is curious that the oldest Holocene Adélie penguin remains so far found elsewhere in this region are less than 1,000 years old. Abandoned breeding sites of Adélie penguins in the Arthur Harbor, Anvers Island, region were occupied 200–600 y.b.p., a period that corresponds to one of environmental cooling known as the Little Ice Age.[18] Presumably, there was more sea ice in the northern Bellingshausen Sea at that time, making the habitat more suitable for this species.[43] In response, Adélie penguins established colonies.

A more complete study of penguin remains in the Antarctic Peninsula region shows that the maximum age of colonies increases with latitude.[17] If penguin remains from 6,000 y.b.p. survived glacial advances in the south (Adelaide Island and vicinity), why would they not survive in the northern portion of the peninsula? Assuming that all remains had an equal chance of surviving glaciation, these results therefore indicate that the oldest colonies in the north are so young only because occupation there has only been recent. In other words, the present colonies in the northern part of the peninsula may be relicts of the Little Ice Age.

Actually, the study of abandoned sites in Arthur Harbor[18] represents dating of an expansion and contraction of the present-day colony, not the colonization and abandonment of separate colonies. This is an example of some confusion resulting from different definitions by different researchers of a penguin colony and a rookery. If we use the definition offered in chapter 3, then the present colony at Arthur Harbor expanded during the Little Ice Age and since has contracted (and is continuing to do so). The same is true for the carbon-dating study of raised beaches in southern Victoria Land.[5,6] Rather than four Adélie penguin colonies having been present in Terra Nova Bay during the penguin optimum (described earlier; fig. 9.10), it appears that the single colony present today had expanded to a much greater size at that time (or, at least, penguins were nesting over a much greater stretch of coastline).

■ Final Thoughts

The Adélie penguin has existed throughout a period of repeated fluctuation between cool glacial and warm interglacial periods. One might surmise that this penguin evolved in such a way as to cope successfully with environmental instability, as long as the fluctuation remained within the limits characteristic of the Pleistocene and Holocene. The species' northern limits are defined (today) by location of the maximum excursion of winter sea ice. Its southern limit is defined by glacial ice of the Antarctic continent itself. As glaciers retreated in the past, the seeming rapidity by which newly exposed terrain was colonized suggests that the extent of philopatry, as discussed earlier, is affected by the availability of nesting habitat. If habitat is available, Adélie penguins apparently can relax what has appeared to be a conservative tendency to emigrate from colonies where they were hatched. Thus the extent of philopatry may not be determined as much by genetically programmed philopatry as was first thought. That is, the tendency to return to the place of birth, long regarded as one of the characteristics that make seabirds what they are, appears not to be an entirely inherited trait or a trait that can be modified easily. That is, the trait is far more plastic in this species (and perhaps others) than it first appears.

If the Adélie penguin, given the chance, is not as philopatric as our short-term modern studies have seemed to indicate, this has implications for any genetic differentiation between populations existing today. Indeed, only two genetic strains have been identified in this penguin thus far.[40,62] In other words, every Adélie penguin is traceable to one of two genetic source populations. In response to the glacial advances of the Pleistocene, what ecologists call the founder effect may have come into play. In that situation, an expansive population today is related to just a few individuals who survived some cataclysm in the past. The advance of ice sheets during glacial periods left little room for Adélie penguins to nest without abandoning their attachment to sea ice-covered seas. However, the species has retained this attachment (ecological trait). During glacial maxima, therefore, very few refugial populations existed. As the glacial conditions relaxed, the species rapidly recolonized its former continental breeding sites. However, this process entailed at least a relaxation of philopatric instincts. Those instincts probably are most evident only when the availability of nesting habitat changes little (i.e., the time span of a modern demographic study; see ch. 8 for further discussion of philopatry).

The dating of penguin remains to chronicle the past behavior of the EAIS and WAIS and to predict how these continental glaciers may respond to environmental change has been an exciting field of investigation. Elsewhere on the planet, bird bones rarely are preserved so well because of the actions of bacteria, mold, and weathering. Equally exciting is the documentation of changes in Adélie penguin populations during the recent years of dramatic warming. In a sense we are living through and observing processes that so far have only been reconstructed through hypothesis to interpret events in the geologic past.

Global warming, an acceleration of the processes that occur naturally as we pass from the LGM, is well under way. Polar regions will be affected most. The rapidly disintegrating ice shelves noted at the beginning of this chapter are tangible evidence of the warming. In addition, processes initiated far in the past, which ultimately may affect Adélie penguins and their marine environment, are also under way. For instance, it has been hypothesized that the retreat of the Ross Ice Shelf (WAIS), under way since the early Holocene (end of the last ice age), will continue regardless of global warming; since then the retreat has proceeded (and may continue) at an average 120 meters per year.[12] In response, ocean currents in the Ross Sea will change and new nesting areas will appear. How a species responds to such a change is no better documented than what the geological, ecological, and genetic record of Adélie penguin colonizations have revealed. The Adélie penguin has shown itself to be a sensitive bellwether and therefore an extremely valuable indicator species of climatic change.

By virtue of the review and synthesis of data contained herein, this volume attests to the huge amount of information that has been amassed about Adélie penguins in only the fifty years since Bill Sladen began his studies at Hope Bay.[42] We know a lot about this polar seabird species, enough that the body of knowledge can and should be a model by which to gauge our understanding of other species in the context of their environments, which likewise are changing. However, much more information on the Adélie penguin is yet to be revealed, and it will be small steps, sometimes repeated, that will carry us toward that end. The easy part is behind us. Future progress will take ever greater effort and creativity on the part of researchers.

Literature Cited

Outline and Chapter 1: Introduction

[1] Ainley, D. G., R. E. LeResche, and W. J. L. Sladen. 1983. *The Breeding Biology of the Adélie Penguin*. Berkeley: University of California Press.

[2] Andersson, K. A. 1905. "Das hohere Tieleben im antarktischen Gebiete." *Vogel Wiss. Ergeb. Schwedischen Sudpolar-Exped., 1901–1903, Vol. Zoologie* 19–58.

[3] Bagshawe, T. W. 1938. "Notes on the Habits of the Gentoo and Ringed or Antarctic Penguins." *Trans. Zool. Soc. London* 24:185–306.

[4] Clarke, W. E. 1913. "Ornithology of the Scottish National Antarctic Expedition: Sect. 5—On the Birds of the South Orkney Islands." *Rept. Sci. Res. Scott. Natl. Antarc. Exped., Zoology* 4:219–247.

[5] Falla, R. A. 1937. "Birds." *B.A.N.Z. Antarc. Res. Exped. 1929–1931, Repts.*, Ser. B, 2:1–288.

[6] Gain, L. 1914. "Oiseaux antarctiques." *Deuxième Expéd. Antarc. Français, 1908–1910* 2:1–200.

[7] Gloersen, P. W., J. Campbell, D. J. Cavalieri, J. C. Comiso, C. L. Parkinson, and H. J. Zwally. 1992. *Arctic and Antarctic Sea Ice, 1978–1987: Satellite Passive-Microwave Observations and Analysis*. NASA SP-511. Washington, D.C.: U.S. National Aeronautics and Space Administration.

[8] Hewitt, R. P. and E. H. Linen Low. 2000. "The Fishery on Antarctic Krill: Defining an Ecosystem Approach to Management." *Rev. Fish. Sci.* 8, no 3:235-298.

[9]Levick, G. M. 1914. *Antarctic Penguins: A Study of Their Social Habits.* New York: McBride, Nast.

[10]Lowe, P. R. and N. B. Kinnear. 1930. "Birds." *British Antarc. ("Terra Nova") Exped., 1910, Nat. Hist. Rept. Zool.* 4, no. 5:103–195.

[11]Matthews, L. H. 1977. *Penguins, Whalers, and Sealers: A Voyage of Discovery.* New York: Universe Books.

[12]Menegaux, A. 1907. *Oiseaux: Expéd. Antarc. Française (1903–1905), commandée par le Dr. Jean Charcot* 4:1–79.

[13]Moss, S. and L. deLeiris. 1988. *Natural History of the Antarctic Peninsula.* New York: Columbia University Press.

[14]Murphy, R. C. 1936. *The Oceanic Birds of South America.* New York: Macmillan.

[15]Neider, C., ed. 1972. *Antarctica: Firsthand Accounts of Exploration and Endurance.* New York: Cooper Square Press.

[16]Richdale, L. E. 1957. *A Population Study of Penguins.* Oxford, U.K.: Oxford University Press.

[17]Roberts, B. B. 1940. "The Breeding Behavior of Penguins with Special Reference to *Pygoscelis papua*." *Brit. Grahamland Exped., 1934–7* 1:195–254.

[18]Rosenman, H., trans. and ed. 1987. *Two Voyages to the South Seas, by Captain J. S.-C. Dumont d'Urville, Volume II: Astrolabe and Zélée, 1837–1840.* Honolulu: University of Hawaii Press.

[19]Ross, J. C. 1847. *A Voyage of Discovery and Research in the Southern and Antarctic Regions During the Years 1939–43.* London: John Murray. Reprint, Devon: David & Charles, 1969.

[20]Sapin-Jaloustre, J. 1960. *Ecologie du Manchot Adélie.* Paris: Hermann.

[21]Shackleton, E. 1909. *The Heart of the Antarctic, the Farthest South Expedition.* London: William Heinemann.

[22]Shackleton, K. 1991. Foreword. In C. Harris and B. Stonehouse, eds., *Antarctica and Global Climate Change,* pp. ix–x. London: Bellhaven Press.

[23]Sladen, W. J. L. 1953. "The Adélie Penguin." *Nature* 171:952–961.

[24]Sladen, W. J. L. 1958. "The Pygoscelid Penguins: I. Methods of Study. II. The Adélie Penguin *Pygoscelis adeliae* (Hombron and Jacquinot)." *Falkland Islands Dependencies Survey, Sci. Repts.,* no. 17.

[25]Sparks, J. and T. Soper. 1967. *Penguins.* Newton Abbot, U.K.: David & Charles.

[26]Stonehouse, B. 1953. "The Emperor Penguin, *Aptenodytes forsteri* (Gray. I.): Breeding Behaviour and Development." *Falkland Islands Dependencies Surveys, Sci. Repts.,* no. 6.

[27]Stonehouse, B. 1956. "The Brown Skua, *Catharacta skua lönnbergi* (Mathews), of South Georgia." *Falkland Islands Dependencies Surveys, Sci. Repts.,* no. 14.

[28]Stonehouse, B. 1960. "The King Penguin, *Aptenodytes patagonica,* of South Georgia, Pt. 1." *Falkland Islands Dependencies Surveys, Sci. Repts.,* no. 23.

[29]Taylor, R. H. 1962. "The Adélie Penguin *Pygoscelis adeliae* at Cape Royds." *Ibis* 104: 176–204.

[30]Wilson, E. A. 1907. "Aves." *Natl. Antarc. Exped. 1901–04,* Vol. 2: *Zoology,* Part 2, pp. 1–121. London: British Museum.

Chapter 2: Marine Ecology

[1]Ainley, D. G. 1980. "Seabirds as Marine Organisms: A Review." *Calif. Co-op. Oceanic Fish. Investig., Rept.* 23:48–53.

[2]Ainley, D. G. and S. S. Jacobs. 1981. "Affinity of Seabirds for Ocean and Ice Boundaries in the Antarctic." *Deep-Sea Res.* 28A:1173–1185.

[3]Ainley, D. G., S. S. Jacobs, C. A. Ribic, and I. Gaffney. 1998. "Seabird Distribution and Oceanic Features of the Amundsen and Southern Bellingshausen Seas." *Antarc. Sci.* 10:111–123.

[4]Ainley, D. G. and R. E. LeResche. 1973. "The Effect of Weather and Ice Conditions on Breeding in Adélie Penguins." *Condor* 75:235–239.

[5]Ainley, D. G., R. E. LeResche, and W. J. L. Sladen. 1983. *Breeding Biology of the Adélie Penguin.* Berkeley: University of California Press.

[6]Ainley, D. G., N. Nur, and E. J. Woehler. 1995. "Factors Affecting the Distribution and Size of Pygoscelid Penguin Colonies in the Antarctic." *Auk* 112:171–182.

[7]Ainley, D. G., E. F. O'Connor, and R. J. Boekelheide. 1984. "The Marine Ecology of Birds in the Ross Sea, Antarctica." *American Ornithol. Union,* Monogr. no. 32.

[8]Ainley, D. G., C. A. Ribic, G. Ballard, P. R. Wilson, and K. R. Barton. 2000. "Foraging-Area Overlap Among Neighboring Colonies of Adélie Penguins: Does Competition Play a Role?" *Abstracts, Fourth International Penguin Conference,* p. 36. Catholic University, Coquimbo, Chile.

[9]Ainley, D. G., C. A. Ribic, and W. R. Fraser. 1992. "Does Prey Preference Affect Habitat Choice in Antarctic Seabirds?" *Mar. Ecol. Progr. Ser.* 90:207–221.

[10]Ainley, D. G., C. A. Ribic, and W. R. Fraser. 1994. "Ecological Structure Among Migrant and Resident Seabirds of the Scotia-Weddell Confluence Region." *J. Anim. Ecol.* 63:347–364.

[11]Ainley, D. G., C. A. Ribic, and L. B. Spear. 1993. "Species-Habitat Relationships Among Antarctic Seabirds: A Function of Physical or Biological Factors." *Condor* 95:806–816.

[12]Ainley, D. G., P. R. Wilson, K. J. Barton, G. Ballard, N. Nur, and B. Karl. 1998. "Diet and Foraging Effort of Adélie Penguins in Relation to Pack-Ice Conditions in the Southern Ross Sea." *Polar Biol.* 20:311–319.

[13]Ashmole, N. P. 1963. "Population Regulation in Seabirds." *Ibis* 103b:458–473.

[14]Baldwin, J. 1988. "Predicting the Swimming and Diving Behaviour of Penguins from Muscle Biochemistry." *Hydrobiologia* 165:255–261.

[15]Baldwin, J., J.-P. Jardel, T. Montague, and R. Tompkin. 1984. "Energy Metabolism in Penguin Swimming Muscles." *Molecular Physiol.* 6:33–42.

[16]Ballard, G., D. G. Ainley, C. A. Ribic, and K. R. Barton. 2001. "Effects of Instrument Attachment and Other Factors on Foraging Trip Duration and Nesting Success of Adélie Penguins." *Condor* 103:481–490.

[17]Bannasch, R. 1995. "Hydrodynamics of Penguins: An Experimental Approach." In P. Dann, I. Norman, and P. Reilly, eds., *The Penguins: Ecology and Management*, pp. 141–176. Chipping North, NSW: Surrey Beatty.

[18]Bierman, W. H. and K. H. Voous. 1950. *Birds Observed and Collected During the Whaling Expeditions of the "Willem Barendsz" in the Antarctic, 1946–1947 and 1947–1948.* Leiden, The Netherlands: E. J. Brill.

[19]Boyd, I. L. and J. P. Croxall. 1996? "Dive Durations in Pinnipeds and Seabirds." *Canad. J. Zool.* 74:1696–1705.

[20]Butler, P. J., R. M. Bevan, A. J. Woakes, and I. L. Boyd. 2000. "Field Metabolic Rate of Gentoo Penguins Throughout the Breeding Season." *Abstracts, Fourth Internatl. Penguin Conference*, p. 18. Catholic University, Coquimbo, Chile.

[21]Chappell, M. A., V. H. Shoemaker, D. N. Janes, T. L. Bucher, and S. K. Maloney. 1993. "Diving Behavior During Foraging in Breeding Adélie Penguins." *Ecology* 74: 1204–1215.

[22]Chappell, M. A., V. H. Shoemaker, D. N. Janes, S. K. Maloney, and T. L. Becher. 1993. "Energetics of Foraging in Breeding Adélie Penguins." *Ecology* 74:2450–2461.

[23]Clarke, J. R. 2001. "Partitioning of Foraging Effort in Adélie Penguins Provisioning Chicks at Béchervaise Island, Antarctica." *Polar Biol.* 24:16–20.

[24]Clarke, J. R., B. Manly, K. Kerry, H. Gardner, E. Franci, S. Corsolini, and S. Focardi. 1998. "Sex Differences in Adélie Penguin Foraging Strategies." *Polar Biol.* 20:248–258.

[25]Cowan, A. N. 1983. "A Modified Penguin Stomach Tube." *Corelia* 7:59–61.

[26]Culik, B. 1992. "Diving Heart Rates in Adélie Penguins *Pygoscelis adeliae*." *Comp. Biochem. Physiol.* A 102:487–490.

[27]Culik, B. 1994. "Energetic Costs of Raising Pygoscelid Penguin Chicks." *Polar Biol.* 14:205–210.

[28]Culik, B. and R. P. Wilson. 1991. "Energetics of Under-Water Swimming in Adélie Penguins (*Pygoscelis adeliae*)." *J. Comp. Physiol.* B 161:285–291.

[29]Culik, B. M., R. P. Wilson, and R. Bannasch. 1994. "Underwater Swimming at Low Energetic Cost by Pygoscelid Penguins." *J. Exp. Biol.* 197:65–78.

[30]Davis, L. S. 1995. "The Control of Behaviour: Free-Running Circadian Rhythms in the Antarctic Summer." In P. Dann, I. Norman, and P. Reilly, eds., *The Penguins: Ecology and Management*, pp. 56–72. Chipping North, NSW: Surrey Beatty.

[31]Davis, L. S., P. D. Boersma, and G. S. Court. 1996. "Satellite Telemetry of the Winter Migration of Adélie Penguins." *Polar Biol.* 16:221–225.

[32]Davis, L. S., G. D. Ward, and R. M. F. S. Sadleir. 1988. "Foraging by Adélie Penguins During the Incubation Period." *Notornis* 35:15–23.

[33]Emison, W. B. 1968. "Feeding Preferences of the Adélie Penguin at Cape Crozier, Ross Island." In O. L. Austin, ed., Antarctic Bird Studies. *Antarc. Res. Ser.* 12:191–212. Washington, D.C.: American Geophysics Union.

[34]Emlen, J. T. and R. L. Penney. 1964. "The Navigation of Penguins." *Nat. Hist.* 73:105–113.

[35] Falla, R. A. 1937. "Birds." *B.A.N.Z. Antarctic Res. Exped. 1929–1931. Repts.*, Series B, vol. II. Adelaide: B.A.N.Z.A.R. Expedition Committee. Adelaide: Hassell Press.

[36] Fraser, W. R. and D. G. Ainley. 1992. *Monitoring Activities at Palmer Station, Antarctica.* Annual Report: AMLR. La Jolla, CA: NOAA, NMFS, Southwest Fisheries Service Science Center.

[37] Fraser, W. R. and W. Z. Trivelpiece. 1996. "Factors Controlling the Distribution of Seabirds: Winter-Summer Heterogeneity in the Distribution of Adélie Penguin Populations." In R. M. Ross, E. E. Hofmann, and L. B. Quetin, eds., Foundations for Ecological Research West of the Antarctic Peninsula. *Antarc. Res. Ser.* 70:257–272. Washington, D.C.: American Geophysics Union.

[38] Furness, R. W. and T. R. Birkhead. 1984. "Seabird Colony Distributions Suggest Competition for Food Supplies During the Breeding Season." *Nature* 311:655–656.

[39] Gloersen, P. W., J. Campbell, D. J. Cavalieri, J. C. Comiso, C. L. Parkinson, and H. J. Zwally. 1992. *Arctic and Antarctic Sea Ice, 1978–1987: Satellite Passive-Microwave Observations and Analysis.* NASA SP-511. Washington, D.C.: U.S. National Aeronautics and Space Administration.

[40] Gurney, A. 1997. *Below the Convergence: Voyages Toward Antarctica 1699–1839.* New York: Penguin.

[41] Hamner, W. M. 1985. "The Importance of Ethology for Investigations of Marine Zooplankton." *Bull. Mar. Sci.* 37:414–424.

[42] Hamner, W. M. and P. P. Hamner. 2000. "Behavior of Antarctic Krill (*Euphausia superba*): Schooling, Foraging, and Antipredator Behavior." *Canad. J. Fish. Aquatic Sci.* 57(Suppl. 3):192–202.

[43] Heil, P. and I. Allison. 1999. "The Pattern and Variability of Antarctic Sea-Ice Drift in the Indian Ocean and Western Pacific Sectors." *J. Geophys. Res.* 104, no. C7: 15789–15802.

[44] Hewitt, R. P. and E. H. Linen Low. 2000. "The Fishery on Antarctic Krill: Defining an Ecosystem Approach to Management." *Rev. Fish. Sci.* 8, no. 3:235–298.

[45] Hill, H. J., P. N. Trathan, J. P. Croxall, and J. L. Watkins. 1996. "A Comparison of Antarctic Krill *Euphausia superba* Caught in Nets and Taken by Macaroni Penguins *Eudyptes chrysolophus*: Evidence for Selection?" *Mar. Ecol. Progr. Ser.* 140:1–11.

[46] Hopkins, T. L. 1987. "Midwater Food Web in McMurdo Sound, Ross Sea, Antarctica." *Mar. Biol.* 96:93–106.

[47] Hosie, G. W. and T. G. Cochran. 1994. "Mesoscale Distribution Patterns of Macrozooplankton Communities in Prydz Bay, Antarctica, January to February 1991." *Mar. Ecol. Prog. Ser.* 106:21–39.

[48] Hunt, G. L., D. Heineman, R. R. Veit, R. B. Heywood, and I. Everson. 1990. "The Distribution, Abundance and Community Structure of Marine Birds in Southern Drake Passage and Bransfield Strait, Antarctica." *Continental Shelf Res.* 10: 243–257.

[49] Ichii, T. 1990. "Distribution of Antarctic Krill Concentrations Exploited by Japanese Krill Trawlers and Minke Whales." *Proc. NIPR Symp. Polar Biol.* 3:36–56.

[50] Janes, D. N. 1997. "Energetics, Growth and Body Composition of Adélie Penguin Chicks." *Physiol. Zool.* 70:237–243.

[51] Joiris, C. R. 1991. "Spring Distribution and Ecological Role of Seabirds and Marine Mammals in the Weddell Sea, Antarctica." *Polar Biol.* 11:415–424.

[52] Kent, S., J. Seddon, G. Robertson, and B. C. Wienecke. 1998. "Diet of Adélie Penguins at Shirley Island, East Antarctica, January 1992." *Mar. Ornithol.* 26:7–10.

[53] Kerry, K. R., J. R. Clarke, and G. D. Else. 1995. "The Foraging Range of Adélie Penguins at Béchervaise Island, MacRobertson Land, Antarctica, as Determined by Satellite Telemetry." In P. Dann, I. Norman, and P. Reilly, eds., *The Penguins: Ecology and Management*, pp. 216–143. Chipping North, NSW: Surrey Beatty.

[54] Knox, G. A. 1994. *The Biology of the Southern Ocean.* Cambridge, U.K.: Cambridge University Press.

[55] Koch, K.-H. and H. H. Reinsch. 1978. "Ornithological Observations During the 'German Antarctic Expedition 1975/76.'" *Beitr. Vogelkd., Leipzig* 24, no. 6:305–328.

[56] Kooyman, G. L. 1975. "Behaviour and Physiology of Diving." In B. Stonehouse, ed., *The Biology of Penguins*, pp. 115–137. London: Macmillan.

[57] Kooyman, G. L. 1989. *Diverse Divers: Physiology and Behavior.* Berlin: Springer-Verlag.

[58] Kooyman, G. L., C. M. Drabek, R. Elsner, and W. B. Campbell. 1971. "Diving Behavior of the Emperor Penguin *Aptenodytes forsteri*." *Auk* 88:775–795.

[59] Kooyman, G. L. and P. J. Ponganis. 1998. "The Physiological Basis of Diving to Depth: Birds and Mammals." *Annu. Rev. Physiol.* 60:19–32.

[60] Levick, G. M. 1914. *Antarctic Penguins: A Study of Their Social Habits.* New York: McBride, Nast.

[61] Levick, G. M. 1915. "Natural History of the Adélie Penguin." *British Antarc. ("Terra Nova") Exped., 1910–1913, Zool.* 1:55–84.

[62] Lishman, G. S. 1985. "The Food and Feeding Ecology of Adélie Penguins (*Pygoscelis adeliae*) and Chinstrap Penguins (*P. antarctica*) at Signy Island, South Orkney Islands." *J. Zool., London* (A) 205:245–263.

[63] Montague, T. L. 1988. "Birds of Prydz Bay, Antarctica: Distribution and Abundance." *Hydrobiologica* 165:227–237.

[64] Moore, J. K., M. R. Abbott, and J. G. Richman. 1999. "Location and Dynamics of the Antarctic Polar Front from Satellite Sea Surface Temperature Data." *J. Geophys. Res.* 104, no. C2:3059–3073.

[65] Müller-Schwarze, D. 1968. "Circadian Rhythms of Activity in the Adélie Penguin (*Pygoscelis adeliae*) During the Austral Summer." In O. L. Austin, ed., *Antarctic Bird Studies. Antarc. Res. Ser.* 12:191–212. Washington, D.C.: American Geophysics Union.

[66] Murphy, R. C. 1936. *The Oceanic Birds of South America.* New York: Macmillan.

[67] Naito, Y., T. Asaga, and Y. Ohyama. 1990. "Diving Behavior of Adélie Penguins Determined by Time-Depth Recorder." *Condor* 92:582–586.

[68] Nicol, S., T. Pauly, N. L. Bindoff, S. Wright, D. Thiele, G. W. Hosle, P. G. Strutton,

and E. Woehler. 2000. "Ocean Circulation Off East Antarctica Affects Ecosystem Structure and Sea-Ice Extent." *Nature* 406:504–507.

[69]Norman, F. I. and S. J. Ward. 1993. "Foraging Group Size and Dive Duration of Adélie Penguins *Pygoscelis adeliae* at Sea Off Hop Island, Rauer Group, East Antarctica." *Mar. Ornithol.* 21:37–47.

[70]Orsii, A. H., T. Witworth III, and W. D. Nowlin, Jr. 1995. "On the Meridional Extent and Fronts of the Antarctic Circumpolar Current." *Deep-Sea Res.* 42:641–673.

[71]Parmalee, D. F., W. R. Fraser, and D. R. Nielson. 1977. "Birds of the Palmer Station Area." *Antarc. J. U.S.* 12, nos. 1 & 2:14–21.

[72]Parrish, J. K. and L. Edelstein-Keshet. 1999. "Complexity, Pattern, and Evolutionary Trade-Offs in Animal Aggregation." *Science* 284:99–101.

[73]Parrish, J. K. and P. Turchin. 1997. "Individual Decisions, Traffic Rules, and Emergent Pattern in Schooling Fish." In J. K. Parrish and W. M. Hamner, eds., *Animal Groups in Three Dimensions,* pp. 126–142. Cambridge, U.K.: Cambridge University Press.

[74]Penney, R. L. 1968. "Territorial and Social Behavior in the Adélie Penguin." In O. L. Austin, ed., Antarctic Bird Studies. *Antarc. Res. Ser.* 12:83–131. Washington, D.C.: American Geophysics Union.

[75]Pennycuick, C. J., J. P. Croxall, and P. A. Prince. 1984. "Scaling of Foraging Radius and Growth Rate in Petrels and Albatrosses (Procellariiformes)." *Ornis Scandinavica* 15:145–154.

[76]Ponganis, P. J. and G. L. Kooyman. 2000. "Diving Physiology of Birds: A History of Studies on Polar Species." *Comp. Biochem. Physiol.* A126:143–151.

[77]Puddicome, R. A. and G. W. Johnstone. 1988. "The Breeding Season Diet of Adélie Penguins at the Vestfold Hills, East Antarctica." *Hydrobiologia* 165:239–253.

[78]Reid, K., P. N. Trathan, J. P. Croxall, and H. J. Hill. 1996. "Krill Caught by Predators and Nets: Differences Between Species and Techniques." *Mar. Ecol. Progr. Ser.* 140:13–20.

[79]Reid, T. A., C. L. Hull, D. W. Eades, R. P. Scofield, and E. J. Woehler. 1999. "Shipboard Observations of Penguins at Sea in the Australian Sector of the Southern Ocean, 1991–1995." *Mar. Ornithol.* 27:101–110.

[80]Ribic, C. A. and D. G. Ainley. 1987/88. "Constancy of Seabird Species Assemblages: An Exploratory Look." *Biolog. Oceanogr.* 6:175–202.

[81]Ribic, C. A., D. G. Ainley, B. J. Karl, and P. R. Wilson. 1998. "Feeding Area Overlap of Adélie Penguin Colonies in the Southern Ross Sea: A Test of the Hinterland Model" [abstract]. *N.Z. Nat. Sci.* 23, Suppl.:159.

[82]Ritz, D. A. 1997. "Costs and Benefits as a Function of Group Size: Experiments on a Swarming Mysid, *Paramesopodopsis rufa* Fenton." In J. K. Parrish and W. M. Hamner, eds., *Animal Groups in Three Dimensions,* pp. 194–206. Cambridge, U.K.: Cambridge University Press.

[83]Ross, J. C. 1847. *A Voyage of Discovery and Research in the Southern and Antarctic Regions During the Years 1939–43.* London: John Murray; Reprint, Devon: David & Charles Reprints, 1969.

[84]Sadleir, R. M. F. S. and K. M. Lay. 1990. "Foraging Movements of Adélie Penguins (*Pygoscelis adeliae*) in McMurdo Sound." In L. S. Davis and J. T. Darby, eds., *Penguin Biology*, pp. 157–179. San Diego: Academic Press.

[85]Salihoglu, B., W. R. Fraser, and E. E. Hofmann. 2001. "Factors Affecting Fledging Weight of Adélie Penguin (*Pygoscelis adeliae*) Chicks: A Modelling Study." *Polar Biol.* 24:328–337.

[86]Sladen, W. J. L. 1958. "The Pygoscelid Penguins. I. Methods of Study. II. The Adélie Penguin *Pygoscelis adeliae* (Hombron and Jacquinot)." *Falkland Islands Dependencies Survey, Sci. Repts.*, no. 17.

[87]Smith, W. O., ed. 1990. *Polar Oceanography*. Orlando, FL: Academic Press.

[88]Smith, W. O. and E. Sakshaug. 1990. "Polar Phytoplankton." In W. O. Smith, ed., *Polar Oceanography*, pp. 477–526. Orlando, FL: Academic Press.

[89]Taylor, R. H. 1962. "Speed of Adélie Penguins on Ice and Snow." *Notornis* 10:111–113.

[90]Thurston, M. H. 1982. "Ornithological Observations in the South Atlantic Ocean and Weddell Sea, 1959–64." *Br. Antarc. Surv. Bull.* 55:77–103.

[91]Trathan, P. N., J. P. Croxall, and E. J. Murphy. 1996. "Dynamics of Antarctic Penguin Populations in Relation to Inter-Annual Variability in Sea Ice Distribution." *Polar Biol.* 16:321–330.

[92]Trivelpiece, W. Z., J. L. Bengtson, S. G. Trivelpiece, and N. J. Volkman. 1986. "Foraging Behavior of Gentoo and Chinstrap Penguins as Determined by New Radio Telemetry Techniques." *Auk* 103:777–781.

[93]Trivelpiece, W. Z., S. G. Trivelpiece, and N. J. Volkman. 1987. "Ecological Segregation of Adélie, Gentoo, and Chinstrap Penguins at King George Island, Antarctica." *Ecology* 68:351–361.

[94]Tynan, C. T. 1998. "Ecological Importance of the Southern Boundary of the Antarctic Circumpolar Current." *Nature* 392:708–710.

[95]van Heezik, Y. 1988. "Diet of Adélie Penguins During the Incubation Period at Cape Bird, Ross Island, Antarctica." *Notornis* 35:23–26.

[96]Volkman, N. J., P. Presler, and W. Trivelpiece. 1980. "Diets of Pygoscelid Penguins at King George Island, Antarctica." *Condor* 82:373–378.

[97]Wanless, S. and M. P. Harris. 1988. "Seabird Records from the Bellingshausen, Amundsen and Ross Seas." *Br. Antarc. Surv. Bull.* 81:87–92.

[98]Watanuki, Y., A. Kato, Y. Mori, and Y. Naito. 1993. "Diving Performance of Adélie Penguins in Relation to Food Availability in Fast Sea-Ice Areas: Comparison Between Years." *J. Anim. Ecol.* 62:634–646.

[99]Watanuki, Y., A. Kato, Y. Naito, G. Robertson, and S. Robinson. 1997. "Diving and Foraging Behaviour of Adélie Penguins in Areas with and without Fast Sea-Ice." *Polar Biol.* 17:296–304.

[100]Watanuki, Y., Y. Mori, and Y. Naito. 1994. "*Euphausia superba* Dominates in the Diet of Adélie Penguins Feeding Under Fast Sea-Ice in the Shelf Areas of Enderby Land in Summer." *Polar Biol.* 14:429–432.

[101] White, M. G. and J. W. H. Conroy. 1975. "Aspects of Competition Between Pygoscelid Penguins at Signy Island, South Orkney Islands." *Ibis* 117:371–373.

[102] Wienecke, B. C., R. Lawless, D. Rodary, C. Bost, R. Thomson, T. Pauly, G. Robertson, K. Kerry, and Y. LeMaho. 2000. "Adélie Penguin Foraging Behaviour and Krill Abundance Along the Wilkes and Adélie Land Coasts, Antarctica." *Deep-Sea Res.* II, 47:2573–2587.

[103] Wilson, E. A. 1907. "Aves." *Natl. Antarctic Exped. 1901–04, Vol. 2. Zoology,* Part 2, pp. 1–121. London: British Museum.

[104] Wilson, R. 1984. "An Improved Stomach Pump for Penguins and Other Seabirds." *J. Field Ornithol.* 55:109–112.

[105] Wilson, R. P. 1995. "Foraging Ecology." In T. D. Williams, ed., *Bird Families of the World, The Penguins Spheniscidae,* pp. 81–106. Oxford, U.K.: Oxford University Press.

[106] Wilson, R. P. and B. M. Culik. 1991. "The Cost of a Hot Meal: Facultative Specific Dynamic Action May Ensure Temperature Homeostasis in Post-Ingestive Endotherms." *Comp. Biochem. Physiol.* 100A:151–154.

[107] Wilson, R. P., B. Culik, D. Adelung, N. Ruben Coria, and H. J. Spairani. 1991. "To Slide or Stride: When Should Adélie Penguins (*Pygoscelis adeliae*) Toboggan?" *Canad. J. Zool.* 69:221–225.

[108] Wilson, R. P., B. M. Culik, D. Adelung, H. J. Spairani, and N. R. Coria. 1991. "Depth Utilisation by Breeding Adélie Penguins, *Pygoscelis adeliae,* at Esperanza Bay, Antarctica." *Mar. Biol.* 109:181–189.

[109] Wilson, R. P., B. Culik, N. R. Coria, D. Adelung, and H. J. Spairani. 1989. "Foraging Rhythms in Adélie Penguins (*Pygoscelis adeliae*) at Hope Bay, Antarctica: Determination and Control." *Polar Biol.* 10:161–165.

[110] Wilson, R. P., B. Culik, P. Korsirek, and D. Adelung. 1998. "The Over-Winter Movements of a Chinstrap Penguin (*Pygoscelis antarctica*)." *Polar Rec.* 34, no. 189:107–112.

[111] Wilson, R. P., K. A. Nagy, and B. S. Obst. 1989. "Foraging Ranges of Penguins." *Polar Rec.* 25, no. 155:303–307.

[112] Wilson, R. P. and G. Peters. 1999. "Foraging Behavior of the Chinstrap Penguin *Pygoscelis antarctica* at Ardley Island, Antarctica." *Mar. Ornithol.* 27:85–95.

[113] Wilson, R. P., K. Peutz, C. A. Bost, B. M. Culik, R. Bannasch, T. Reins, and D. Adelung. 1993. "Diel Dive Depth in Penguins in Relation to Diel Vertical Migration of Prey: Whose Dinner by Candlelight?" *Mar. Ecol. Progr. Ser.* 94:101–104.

[114] Wilson, R. P. and M.-P. T. Wilson. 1990. "Foraging Ecology of Breeding *Spheniscus* Penguins." In L. S. Davis and J. T. Darby, eds., *Penguin Biology,* pp. 181–206. San Diego: Academic Press.

[115] Woehler, E. J. 1995. "Consumption of Southern Ocean Marine Resources by Penguins." In P. Dann, I. Norman, and P. Reilly, eds., *The Penguins: Ecology and Management,* pp. 266–296. Chipping North, NSW: Surrey Beatty.

[116] Woehler, E. J., C. L. Hodges, and D. J. Watts. 1990. "An Atlas of the Pelagic Distri-

bution and Abundance of Seabirds in the Southern Indian Ocean, 1981–1990." *Austr. Natl. Antarc. Res. Expeditions, Res. Notes* 77:1–406.

[117]Zusi, L. 1975. "An Interpretation of Skull Structure in Penguins." In B. Stonehouse, ed., *The Biology of Penguins,* pp. 59–84. London: Macmillan.

[118]Zwally, H. J., J. C. Comiso, C. L. Parkinson, W. J. Campbell, W. D. Casey, and P. Gloersen. 1983. *Antarctic Sea Ice, 1973–1976: Satellite Passive-Microwave Observations.* Washington, D.C.: U.S. National Aeronautics and Space Administration.

Chapter 3: Breeding Populations: Size and Distribution

[1]Ainley, D. G. and D. P. DeMaster. 1980. "Survival and Mortality in a Population of Adélie Penguins." *Ecology* 61:522–530.

[2]Ainley, D. G. and R. E. LeResche. 1973. "The Effects of Weather and Ice Conditions on Breeding in Adélie Penguins." *Condor* 75:235–255.

[3]Ainley, D. G., R. E. LeResche, and W. J. L. Sladen. 1983. *Breeding Biology of the Adélie Penguin.* Berkeley: University of California Press.

[4]Ainley, D. G., S. H. Morrell, and R. C. Wood. 1986. "South Polar Skua Breeding Colonies in the Ross Sea Region, Antarctica." *Notornis* 33:155–163.

[5]Ainley, D. G., N. Nur, and E. C. Woehler. 1995. "Factors Affecting the Distribution and Size of Pygoscelid Penguin Colonies in the Antarctic." *Auk* 112:171–182.

[6]Ashmole, N. P. 1963. "The Regulation of Numbers of Tropical Oceanic Birds." *Ibis* 103b:458–473.

[7]Baroni, C. and G. Orombelli. 1991. "Holocene Raised Beaches at Terra Nova Bay, Victoria Land, Antarctica." *Quaternary Res.* 36:157–177.

[8]Baroni, C. and G. Orombelli. 1994. "Abandoned Penguin Rookeries as Holocene Paleoclimatic Indicators in Antarctica." *Geology* 22:23–26.

[9]Birt, V. L., T. P. Birt, D. Goulet, D. K. Cairns, and W. A. Montevecchi. 1987. "Ashmole's Halo: Direct Evidence for Prey Depletion by a Seabird." *Mar. Ecol. Progr. Ser.* 40:205–208.

[10]Cairns, D. K. 1989. "The Regulation of Seabird Colony Size: A Hinterland Model." *Amer. Natural.* 134:141–146.

[11]Carsey, F. D. 1992. "Microwave Remote Sensing of Sea Ice." *Geophys. Monogr.* 68. Washington, D.C.: American Geophysics Union.

[12]Croxall, J. P., D. M. Rootes, and P. A. Prince. 1981. "Increases in Penguin Populations at Signy Island, South Orkney Islands." *Br. Antarc. Surv. Bull.* 54:47–56.

[13]Emslie, S. D. 1995. "Age and Taphonomy of Abandoned Penguin Rookeries in the Antarctic Peninsula Region." *Polar Rec.* 31, no. 179:409–418.

[14]Falla, R. A. 1937. "Birds." *B.A.N.Z. Antarctic Research Expedition 1929–1931. Repts.,* Series B, vol. 2. Adelaide: B.A.N.Z.A.R. Expedition Committee. Adelaide: Hassell Press.

[15]Fraser, W. R., W. Z. Trivelpiece, D. G. Ainley, and S. G. Trivelpiece. 1992. "Increases in Antarctic Penguin Populations: Reduced Competition with Whales or a Loss of Sea Ice Due to Environmental Warming?" *Polar Biol.* 11:525–531.

[16] Furness, R. W. and T. R. Birkhead. 1984. "Seabird Colony Distributions Suggest Competition for Food Supplies During the Breeding Season." *Nature* 311:655–656.

[17] Gloersen, P., W. J. Campbell, D. J. Cavalieri, J. C. Comiso, C. L. Parkinson, and H. J. Zwally. 1992. *Arctic and Antarctic Sea Ice, 1978–1987: Satellite Passive-Microwave Observations and Analysis*. NASA SP-511. Washington, D.C.: U.S. National Aeronautics and Space Administration.

[18] Holdgate, M. W. 1963. "Observations of Birds and Seals at Anvers Island, Palmer Archipelago, in 1955–57." *Br. Antarc. Surv. Bull.* 2:45–51.

[19] Hoshiai, T., T. Sweda, and A. Tanimura. 1984. "Adélie Penguin Census in the 1981–82 and 1982–83 Breeding Seasons Near Syowa Station, Antarctica." *Mem. Natl. Instit. Polar Res.*, Spec. Issue no. 32:117–121.

[20] Jacobs, S. S., ed. 1985. "Oceanology of the Antarctic Continental Shelf." *Antarc. Res. Ser.* 43. Washington, D.C.: American Geophysics Union.

[21] Jacobs, S. S. and J. C. Comiso. 1989. "Sea Ice and Oceanic Processes on the Ross Sea Continental Shelf." *J. Geophys. Res.* 94, no. C12:18195–18211.

[22] Levick, G. M. 1914. *Antarctic Penguins: A Study of Their Social Habits*. London: William Heinemann.

[23] Levick, G. M. 1915. "Natural History of the Adélie Penguin." *British Antarctic ("Terra Nova") Exped., 1910–1913, Zool.* 1:55–84.

[24] Markus, T., C. Kittmeier, and E. Fahrbach. 1998. "Ice Formation in Coastal Polynyas in the Weddell Sea and Their Impact on Oceanic Salinity." In M. O. Jeffries, ed., "Antarctic Sea Ice: Physical Properties, Interactions and Variability." *Geophys. Monogr.* 74:273–292. Washington, D.C.: American Geophysics Union.

[25] Martin, S., K. Steffen, J. Comiso, D. Cavalieri, M. R. Drinkwater, and B. Holt. 1992. "Microwave Remote Sensing of Polynyas." In F. D. Carsey, ed., "Microwave Remote Sensing of Sea Ice." *Geophys. Monogr.* 68:303–311. Washington, D.C.: American Geophysics Union.

[26] Massom, R. A., P. T. Harris, K. J. Michael, and M. J. Potter. 1998. "The Distribution and Formative Processes of Latent-Heat Polynyas in East Antarctica." *Annals Glaciol.* 27:420–426.

[27] Naganobu, M., K. Kutsuwada, Y. Sasai, S. Taguchi, and V. Siegel. 1999. "Relationships Between Antarctic Krill (*Euphausia superba*) Variability and Westerly Fluctuations and Ozone Depletion in the Antarctic Peninsula Area." *J. Geophys. Res.* 104, no. C9:20651–20665.

[28] Parmelee, D. F., W. R. Fraser, and D. R. Neilson. 1977. "Birds of the Palmer Station Area." *Antarc. J. U.S.* 12, nos. 1 and 2:14–21.

[29] Penney, R. L. 1968. "Territorial and Social Behavior in the Adélie Penguin." In O. L. Austin, ed., "Antarctic Bird Studies." *Antarc. Res. Ser.* 12:83–131. Washington, D.C.: American Geophysics Union.

[30] Prézelin, B. B., E. E. Hofmann, C. Mengelt, and J. M. Klinck. 2000. "The Linkage Between Upper Circumpolar Deep Water (UCDW) and Phytoplankton Assemblages on the West Antarctic Peninsula Continental Shelf." *J. Mar. Res.* 58:165–202.

[31] Reid, B. 1961. "An Assessment of the Size of the Cape Adare Adélie Penguin Rookery and Skuary—with Notes on Petrels." *Notornis* 10:98–111.

[32] Reilly, P. 1983. *Fairy Penguins and Earthy People.* Melbourne: Lothian.

[33] SCAR (Subcommittee on Bird Biology). 1996. *The Status and Trends of Antarctic and Subantarctic Seabirds.* SC-CAMLR-XV/BG/29. Hobart, Tasmania: Commission for the Conservation of Antarctic Marine Living Resources.

[34] SCAR (Subcommittee on Bird Biology; E. J. Woehler, compiler). 2000. *Status and Trends of Antarctic and Subantarctic Penguins, 2000.* Kingston, Tasmania: Australian Antarctic Division.

[35] Sladen, W. J. L. 1958. "The Pygoscelid Penguins. I. Methods of Study. II. The Adélie Penguin *Pygoscelis Adeliae* (Hombron and Jacquinot)." *Falkland Islands Dependencies Survey, Sci. Repts.*, no. 17.

[36] Smith, R. C., D. Ainley, K. Baker, E. Domack, S. Emslie, B. Fraser, J. Kennett, A. Leventer, E. Mosley-Thompson, S. Stammerjohn, and M. Vernet. 1999. "Marine Ecosystem Sensitivity to Climate Change." *BioScience* 49, no. 5:393–404.

[37] Splettstoesser, J. and F. S. Todd. 1999. "Stomach Stones from Emperor Penguin *Aptenodytes forsteri* Colonies in the Weddell Sea." *Mar. Ornithol.* 27:97–100.

[38] Stirling, I. and H. Cleator. 1981. "Polynyas in the Canadian Arctic." *Canad. Wildl. Serv., Occ. Paper* no. 45.

[39] Stonehouse, B. 1964. "Emperor Penguins at Cape Crozier." *Nature* 203, no. 4947:849–851.

[40] Stonehouse, B. 1964. "Bird life." In R. E. Priestley, R. J. Adie, and G. de Q. Robin, eds., *Antarctic Research: A Review of British Scientific Achievement in Antarctica,* p. 229. London: Butterworth.

[41] Stonehouse, B. 1967. "Occurrence and Effects of Open Water in McMurdo Sound, Antarctica, During Winter and Early Spring." *Polar Rec.* 13:775–778.

[42] Stonehouse, B. 1969. "Air Census of Two Colonies of Adélie Penguins (*Pygoscelis adeliae*) in Ross Dependency, Antarctica." *Polar Rec.* 14, no. 91:471–475.

[43] Sturman, A. P. and M. R. Anderson. 1986. "On the Sea-Ice Regime of the Ross Sea, Antarctica." *J. Glaciol.* 32, no. 110:54–59.

[44] Taylor, R. H., P. R. Wilson, and B. W. Thomas. 1990. "Status and Trends of Adélie Penguin Populations in the Ross Sea Region." *Polar Rec.* 26, no. 159:293–304.

[45] Trathan, P. N., J. P. Croxall, and E. J. Murphy. 1996. "Dynamics of Antarctic Penguin Populations in Relation to Inter-Annual Variability in Sea Ice Distribution." *Polar Biol.* 16:321–330.

[46] Trivelpiece, W. Z. and W. R. Fraser. 1996. "The Breeding Biology and Distribution of Adélie Penguins: Adaptations to Environmental Variability." In R. M. Ross, E. E. Hofmann, and L. B. Quetin, eds., Foundations for Ecological Research West of the Antarctic Peninsula. *Antarc. Res. Ser.* 70:273–285. Washington, D.C.: American Geophysics Union.

[47] Trivelpiece, W. Z., S. G. Trivelpiece, G. R. Geupel, J. Kjelmyr, and N. J. Volkman. 1990. "Adélie and Chinstrap Penguins: Their Potential as Monitors of the Southern

Ocean Marine Ecosystem." In K. R. Kerry and G. Hempel, eds., *Ecological Change and the Conservation of Antarctic Ecosystems,* pp. 191–202. Berlin: Springer-Verlag.

[48]Van Woert, M. L. 1999. "Wintertime Dynamics of the Terra Nova Bay Polynya." *J. Geophys. Res.* 104, no. 4:7753–7769.

[49]Volkman, N. J. and W. Trivelpiece. 1981. "Nest-Site Selection Among Adélie, Chinstrap and Gentoo Penguins in Mixed Species Rookeries." *Wilson Bull.* 93:243–248.

[50]Watanuki, Y., A. Kato, Y. Mori, and Y. Naito. 1993. "Diving Performance of Adélie Penguins in Relation to Food Availability in Fast Sea-Ice Areas: Comparison Between Years." *J. Anim. Ecol.* 62:634–646.

[51]Watanuki, Y., Y. Mori, and Y. Naito. 1994. "*Euphausia superba* Dominates in the Diet of Adélie Penguins Feeding Under Fast Sea-Ice in the Shelf Areas of Enderby Land in Summer." *Polar Biol.* 14:429–432.

[52]White, M. G. and J. W. H. Conroy. 1975. "Aspects of Competition Between Pygoscelid Penguins at Signy Island, South Orkney Islands." *Ibis* 117:371–373.

[53]Whitehead, M. C. and G. W. Johnstone. 1990. "The Distribution and Estimated Abundance of Adélie Penguins Breeding in Prydz Bay, Antarctica." *Proc. Natl. Instit. Polar Res., Symp. Polar Biol.* 3:91–98.

[54]Wilson, K.-J. 1990. "Fluctuations of Populations of Adélie Penguins at Cape Bird, Antarctica." *Polar Rec.* 26, no. 159:305–308.

[55]Woehler, E. J. 1993. *The Distribution and Abundance of Antarctic and Subantarctic Penguins.* Cambridge, U.K.: Scientific Committee for Antarctic Research, Scott Polar Research Institute.

[56]Woehler, E. J. 1995. "Consumption of Southern Ocean Marine Resources by Penguins." In P. Dann, I. Norman, and P. Reilly, eds., *The Penguins: Ecology and Management,* pp. 266–295. Chipping North, NSW: Surrey Beatty.

[57]Woehler, E. J., D. J. Slip, L. M. Robertson, P. J. Fullagar, and H. R. Burton. 1991. "The Distribution, Abundance and Status of Adélie Penguins *Pygoscelis adeliae* at the Windmill Islands, Wilkes Land, Antarctica." *Mar. Ornithol.* 19:1–18.

[58]Worby, A. P., R. A. Massom, I. Allison, V. I. Lytle, and P. Heil. 1998. "East Antarctic Sea Ice: A Review of Its Structure, Properties and Drift." In M. O. Jeffries, ed., Antarctic Sea Ice: Physical Properties, Interactions and Variability. *Geophys. Monogr.* 74:41–67. Washington, D.C.: American Geophysics Union.

[59]Zwally, H. J., J. C. Comiso, C. L. Parkinson, W. J. Campbell, W. D. Casey, and P. Gloersen. 1983. *Antarctic Sea Ice, 1973–1976: Satellite Passive-Microwave Observations.* Washington, D.C.: U.S. National Aeronautics Space Administration.

Chapter 4: The Annual Cycle

[1]Ainley, D. G. 1975. "The Development of Reproductive Maturity in Adélie Penguins." In B. Stonehouse, ed., *The Biology of Penguins,* pp. 139–157. London: Macmillan.

[2]Ainley, D. G. 1978. "Activity of Non-Breeding Adélie Penguins." *Condor* 80:135–146.

[3]Ainley, D. G. and W. B. Emison. 1972. "Sexual Size Dimorphism in Adélie Penguins." *Ibis* 114:267–271.

[4]Ainley, D. G., S. S. Jacobs, C. A. Ribic, and I. Gaffney. 1998. "Seabird Distribution and Oceanic Features of the Amundsen and Southern Bellingshausen Seas." *Antarc. Sci.* 10:111–123.

[5]Ainley, D. G. and R. E. LeResche. 1973. "The Effect of Weather and Ice Conditions on Breeding in Adélie Penguins." *Condor* 75:235–239.

[6]Ainley, D. G., L. E. LeResche, and W. J. L. Sladen. 1983. *Breeding Biology of the Adélie Penguin.* Berkeley: University of California Press.

[7]Ainley, D. G., E. F. O'Connor, and R. J. Boekelheide. 1984. "The Marine Ecology of Birds in the Ross Sea, Antarctica." *Amer. Ornithol. Union Monogr.* no. 32.

[8]Ainley, D. G., C. A. Ribic, and R. C. Wood. 1990. "A Demographic Study of the South Polar Skua at Cape Crozier." *J. Anim. Ecol.* 59:1–20.

[9]Ainley, D. G., R. C. Wood, and W. J. L. Sladen. 1978. "Bird Life at Cape Crozier, Ross Island." *Wilson Bull.* 90:492–510.

[10]Amadon, D. 1964. "The Evolution of Low Reproductive Rates in Birds." *Evolution* 18:105–110.

[11]Ashmole, N. P. and H. Tovar. 1968. "Prolonged Parental Care in Royal Terns and Other Birds." *Auk* 85:90–100.

[12]Austin, O. L., Jr. 1957. "Notes on Banding Birds in Antarctica, and on the Adélie Penguin Colonies of the Ross Sea Sector." *Bird-Banding* 28:1–26.

[13]Bagshawe, T. W. 1938. "Notes on the Habits of the Gentoo and Ringed or Antarctic Penguins." *Trans. Zool. Soc. London* 24, no. 3:185–307.

[14]Boersma, D. B. 1978. "Breeding Patterns of Galápagos Penguins as an Indicator of Oceanographic Conditions." *Science* 200:1481–1483.

[15]Buckley, F. G. and P. A. Buckley. 1974. "Comparative Feeding Ecology of Wintering Adult and Juvenile Royal Terns (Aves: Laridae, Sterninae)." *Ecology* 55:1053–1063.

[16]Carrick, R. 1970. "Ecology and Population Dynamics of Antarctic Sea-Birds." In M. Holdgate, ed., *Antarctic Ecology,* pp. 505–525. New York: Academic Press.

[17]Carrick, R. 1972. "Population Ecology of the Australian Black-Backed Magpie, Royal Penguin, and Silver Gull." In Population Ecology of Migratory Birds. *U.S. Dept. Interior, Wildl. Res. Rept.* 2:41–99.

[18]Carrick, R. and S. E. Ingham. 1967. "Antarctic Sea-Birds as Subjects for Ecological Research." *J.A.R.E. Sci. Rept.,* Spec. Issue no. 1:151–184.

[19]Cendron, J. 1953. "La Mue du Manchot Adélie Adulte." *Alauda* 21:77–84.

[20]Coulson, J. C. and E. White. 1958. "The Effect of Age on the Breeding Biology of the Kittiwake, *Rissa tridactyla.*" *Ibis* 100:40–51.

[21]Dunn, E. K. 1972. "Effect of Age on the Fishing Ability of Sandwich Terns, *Sterna sandwichensis.*" *Ibis* 114:360–366.

[22]Ficken, M. S. and R. W. Ficken. 1967. "Age-Specific Differences in the Breeding Behavior and Ecology of the American Redstart." *Wilson Bull.* 79:188–198.

[23]Hanson, N. 1902. *Extracts from the Private Diary of the Late Nicolai Hanson. Report on the Collections of Natural History Made in the Antarctic Regions During the Voyage of the "Southern Cross,"* Pt. 3, pp. 79–105. London: British Museum.

[24] Harrington, B. A. 1974. "Colony Visitation Behavior and Breeding Ages of Sooty Terns (*Sterna fuscata*)." *Bird-Banding* 45:115–144.

[25] Harris, M. P. 1966. "Age of Return to the Colony, Age of Breeding and Adult Survival of Manx Shearwaters." *Bird Study* 13:84–95.

[26] Kerry, K., J. Clarke, and G. Else. 1993. "The Use of an Automated Weighing and Recording System for the Study of the Biology of Adélie Penguins (*Pygoscelis adeliae*)." *Proc. NIPR Symp Polar Biol.* 6:62–75.

[27] Lack, D. 1966. *Population Studies of Birds.* Oxford, U.K.: Clarendon Press.

[28] Lack, D. 1968. *Ecological Adaptations for Breeding in Birds.* London: Methuen.

[29] LeResche, R. E. 1971. *Ecology and Behavior of Known-Age Adélie Penguins.* PhD dissertation, Johns Hopkins University, Baltimore, MD.

[30] Levick, G. M. 1915. "Natural History of the Adélie Penguin." *British Antarc. ("Terra Nova") Expedition, 1910, Zoology* 1:55–84.

[31] Lishman, G. S. 1985. "The Comparative Breeding Biology of Adélie and Chinstrap Penguins *Pygoscelis adeliae* and *P. antarctica* at Signy Island, South Orkney Islands." *Ibis* 127:84–99.

[32] Murton, R. K. and N. J. Westwood. 1977. *Avian Breeding Cycles.* Oxford, U.K.: Clarendon Press.

[33] Nelson, J. B. 1966. "The Behaviour of the Young Gannet." *British Birds* 59:393–419.

[34] Nelson, J. B. 1978. *The Gannet.* Vermillion, SD: Buteo Books.

[35] Orians, G. H. 1969. "Age and Hunting Success in the Brown Pelican (*Pelecanus occidentalis*)." *Anim. Beh.* 17:316–319.

[36] Parmalee, D. F., W. R. Fraser, and D. R. Nielson. 1977. "Birds of the Palmer Station Area." *Antarc. J. U.S.* 12, nos. 1 and 2:14–21.

[37] Penney, R. L. 1967. "Molt in the Adélie Penguin." *Auk* 84:61–71.

[38] Penney, R. L. 1968. "Territorial and Social Behavior in the Adélie Penguin." In A. O. Austin, Jr., ed., Antarctic Bird Studies. *Antarc. Res. Ser.* 12:83–132. Washington, D.C.: American Geophysics Union.

[39] Recher, H. G. and J. A. Recher. 1969. "Comparative Foraging Efficiency of Adult and Immature Little Blue Herons (*Florida caerulea*)." *Anim. Beh.* 17:320–322.

[40] Reid, B. 1964. "The Cape Hallett Adélie Penguin Rookery (Its Size, Composition, and Structure)." *Rec. Dominion Mus., Wellington* 5:11–37.

[41] Rice, D. W. and K. W. Kenyon. 1962. "Breeding Cycles and Behavior of Laysan and Black-Footed Albatrosses." *Auk* 79:517–567.

[42] Richdale. L. E. 1957. *A Population Study of Penguins.* Oxford, U.K.: Oxford University Press.

[43] Sapin-Jaloustre, J. 1960. "Ecologie du Manchot Adélie." *Expeditions Polaires Française,* no. 208. Paris: Hermann.

[44] Serventy, D. L. and P. J. Curry. 1984. "Observations on Colony Size, Breeding Success, Recruitment and Inter-Colony Dispersal in a Tasmanian Colony of Short-Tailed Shearwaters *Puffinus tenuirostris* over a 30-Year Period." *Emu* 84:71–79.

[45] Sladen, W. J. L. 1958. "The Pygoscelid Penguins. I. Methods of Study. II. The Adélie

Penguin *Pygoscelis adeliae* (Hombron and Jacquinot)." *Falkland Islands Dependencies Survey, Sci. Repts.*, no. 17.

[46]Smith, W. B. 1990. *Polar Oceanography.* San Diego: Academic Press.

[47]Spurr, E. B. 1975. "Breeding of the Adélie Penguin *Pygoscelis adeliae* at Cape Bird." *Ibis* 117:324–338.

[48]Taylor, R. H. 1962. "The Adélie Penguin *Pygoscelis adeliae* at Cape Royds." *Ibis* 104:176–204.

[49]Taylor, R. H., P. R. Wilson, and B. W. Thomas. 1990. "Status and Trends of Adélie Penguin Populations in the Ross Sea Region." *Polar Rec.* 26, no. 159:293–304.

[50]Van Tienhoven, A. 1961. "Endocrinology of Reproduction in Birds." In W. C. Young, ed., *Sex and Internal Secretions,* Vol. II, 3rd ed., pp. 1088–1172. Baltimore, MD: Williams & Wilkins.

[51]Waas, J. R. 1995. "Social Stimulation and Reproductive Schedules: Does the Acoustic Environment Influence the Egg-Laying Schedule in Penguin Colonies?" In P. Dann, I. Norman, and P. Reilly, eds., *The Penguins: Ecology and Management,* pp. 111–137. Chipping North, NSW: Surrey Beatty.

[52]Ward, R. and A. Zahavi. 1973. "The Importance of Certain Assemblages of Birds as 'Information Centres' for Food Finding." *Ibis* 115:517–534.

[53]Welty, J. C. 1962. *The Life of Birds.* Philadelphia: W. B. Saunders.

[54]Whitehead, M. D., G. W. Johnstone, and H. R. Burton. 1990. "Annual Fluctuations in Productivity and Breeding Success of Adélie Penguins and Fulmarine Petrels in Prydz Bay, East Antarctica." In K. R. Kerry and G. Hempel, eds., *Antarctic Ecosystems: Ecological Change and Conservation,* pp. 214–223. Berlin: Springer-Verlag.

[55]Wilson, E. A. 1907. "Aves." *Natl. Antarctic Exped. 1901–04, Vol. 2. Zool.* 2:1–121.

[56]Wilson, P. R., D. G. Ainley, N. Nur, S. S. Jacobs, K. J. Barton, G. Ballard, and J. C. Comiso. 2001. "Adélie Penguin Population Change in the Pacific Sector of Antarctica: Relation to Sea-Ice Extent and the Antarctic Circumpolar Current." *Mar. Ecol. Progr. Ser.* 213:301–309.

[57]Wooler, R. D. and J. C. Coulson. 1977. "Factors Affecting the Age of First Breeding of the Kittiwake, *Rissa tridactyla.*" *Ibis* 119:339–349.

[58]Yeates, G. W. 1968. "Studies on the Adélie Penguin at Cape Royds 1964–65 and 1965–66." *N.Z. J. Mar. Freshwater Res.* 2:472–496.

Chapter 5: The Occupation Period: Pair Formation, Egg Laying, and Incubation

[1]Ainley, D. G. 1975. "Displays of Adélie Penguins: A Reinterpretation." In B. Stonehouse, ed., *The Biology of Penguins,* pp. 503–534. London: Macmillan.

[2]Ainley, D. G. 1978. "Activity of Non-Breeding Adélie Penguins." *Condor* 80:135–146.

[3]Ainley, D. G. and D. P. DeMaster. 1980. "Survival and Mortality in a Population of Adélie Penguins." *Ecology* 61:522–530.

[4]Ainley, D. G. and R. E. LeResche. 1973. "The Effects of Weather and Ice Conditions on Breeding in Adélie Penguins." *Condor* 75:235–255.

[5] Ainley, D. G., R. E. LeResche, and W. J. L. Sladen. 1983. *Breeding Biology of the Adélie Penguin*. Berkeley: University of California Press.

[6] Ainley, D. G., C. A. Ribic, and R. C. Wood. 1990. "A Demographic Study of the South Polar Skua at Cape Crozier." *J. Anim. Ecol.* 59:1–20.

[7] Astheimer, L. B. and C. R. Grau. 1985. "The Timing and Energetic Consequences of Egg Formation in the Adélie Penguin." *Condor* 87:256–268.

[8] Bagshawe, T. W. 1938. "Notes on the Habits of the Gentoo and Ringed or Antarctic Penguins." *Trans. Zool. Soc. London* 24:185–306.

[9] Coulson, J. C. and E. White. 1960. "The Effect of Age and Density of Breeding Birds on the Time of Breeding of the Kittiwake, *Rissa tridactyla*." *Ibis* 102:71–86.

[10] Croxall, J. P., T. S. McCann, P. A. Prince, and P. Rothery. 1988. "Reproductive Performance of Seabirds and Seals at South Georgia and Signy Island, South Orkney Islands, 1976–1987: Implications for Southern Ocean Monitoring Studies." In D. Sahrhage, ed., *Antarctic Ocean and Resources Variability*, pp. 261–285. Berlin: Springer-Verlag.

[11] Davis, L. S. and F. T. McCaffrey. 1986. "Survival Analysis of Eggs and Chicks of Adélie Penguins (*Pygoscelis adeliae*)." *Auk* 103:379–388.

[12] Fraser, W. R. and D. L. Patterson. 1997. "Human Disturbance and Long-Term Changes in Adélie Penguin Populations: A Natural Experiment at Palmer Station, Antarctica." In B. Bataglia, J. Valencia, and D. W. H. Walton, eds., *Antarctic Communities: Species, Structure and Survival*, pp. 445–452. Cambridge, U.K.: Cambridge University Press.

[13] Goldsmith, R. and W. J. L. Sladen. 1961. "Temperature Regulation of Some Antarctic Penguins." *J. Physiol.* 157:251–262.

[14] Grau, C. R. 1984. "Egg Formation." In G. C. Whittow and H. Rahn, eds., *Seabird Energetics*, pp. 33–58. New York: Plenum Press.

[15] Harris, M. P. 1966. "The Breeding Biology of the Manx Shearwater." *Ibis* 108:17–33.

[16] Irvine, L. G., J. R. Clarke, and K. R. Kerry. 2000. "Poor Breeding Success of the Adélie Penguin at Béchervaise Island in the 1998–99 Season." *CCAMLR Science* 7:151–167.

[17] Kerry, K., J. Clarke, and G. Else. 1993. "The Use of an Automated Weighing and Recording System for the Study of the Biology of Adélie Penguins (*Pygoscelis adeliae*)." *Proc. NIPR Symp. Polar Biol.* 6:62–75.

[18] LeResche, R. E. 1971. *Ecology and Behavior of Known-Age Adélie Penguins*. Ph.D. dissertation, Johns Hopkins University, Baltimore, MD.

[19] LeResche, R. E. and W. J. L. Sladen. 1970. "Establishment of Pair and Breeding Site Bonds by Known-Age Adélie Penguins." *Anim. Beh.* 18:517–526.

[20] Levick, G. M. 1914. *Antarctic Penguins: A Study of Their Social Habits*. New York: McBride, Nast.

[21] Lishman, G. S. 1985. "The Comparative Breeding Biology of Adélie and Chinstrap Penguins *Pygoscelis adeliae* and *P. antarctica* at Signy Island, South Orkney Islands." *Ibis* 127:84–99.

[22] Maher, W. J. 1966. "Predation Impact on Penguins." *Natural History* 75:42–51.

[23] Mills, J. A. 1973. "The Influence of Age and the Pair-Bond on Breeding of the Red-Billed Gull, *Larus novaehollandiae scolopinus*." *J. Anim. Ecol.* 42:147–162.

[24] Moczydlowski, E. 1986. "Microclimate of the Nest-Sites of Pygoscelid Penguins (Admiralty Bay, South Shetland Island)." *Polish Polar Res.* 7, no. 4:377–394.

[25] Mougin, J.-L. 1968. "Notes sur le Cycle Reproducteur et la Mue du Manchot Adélie (*Pygoscelis adeliae*) dans l'Archipel de Point Géologie (Terre Adélie)." *L'Oiseau et R.F.O.* 38:89–94.

[26] Müller-Schwarze, C. and D. Müller-Schwarze. 1975. "A Survey of Twenty-Four Rookeries of Pygoscelid Penguins in the Antarctic Peninsula Region." In B. Stonehouse, ed., *The Biology of Penguins*, pp. 307–320. London: Macmillan.

[27] Penney, R. L. 1968. "Territorial and Social Behavior in the Adélie Penguin." In O. A. Austin, Jr., ed., Antarctic Bird Studies. *Antarc. Res. Ser.* 12:83–132. Washington, D.C.: American Geophysical Union.

[28] Preston, W. 1968. *Breeding Ecology and Social Behavior of the Black Guillemot*, Cepphus grylle. Ph.D. dissertation, University of Michigan, Ann Arbor.

[29] Reid, B. 1964. "The Cape Hallett Adélie Penguin Rookery (Its Size, Composition, and Structure)." *Rec. Dominion Mus., Wellington* 5:11–37.

[30] Reid, B. 1965. "The Adélie Penguin (*Pygoscelis adeliae*) Egg." *N.Z. J. Sci.* 8:503–514.

[31] Reid, B. E. and C. Bailey. 1966. "The Value of the Yolk Reserve in Adélie Penguin Chicks." *Rec. Dominion Mus., Wellington* 5, no. 19:185–193.

[32] Richdale, L. E. 1954. "Breeding Efficiency in the Yellow-Eyed Penguin." *Ibis* 96: 207–224.

[33] Richdale, L. E. 1957. *A Population Study of Penguins*. Oxford, U.K.: Oxford University Press.

[34] Romanoff, A. L. and A. J. Romanoff. 1949. *The Avian Egg*. New York: Wiley.

[35] Roudybush, T. E., C. R. Grau, M. R. Petersen, D. G. Ainley, K. V. Hirsch, A. P. Gilman, and S. M. Patten. 1979. "Yolk Formation in Some Charadriiforme Birds." *Condor* 81:293–298.

[36] Sapin-Jaloustre, J. 1960. "Ecologie du Manchot Adélie." *Expeditions Polaires Française*, no. 208. Paris: Hermann.

[37] Serventy, D. L. 1963. "Egg Laying Time Table of the Slender-Billed Shearwater, *Puffinus tenuirostris*." *Proc. Internatl. Ornithol. Congr.* 13:338–343.

[38] Sladen, W. J. L. 1958. "The Pygoscelid Penguins. I. Methods of Study. II. The Adélie Penguin *Pygoscelis adeliae* (Hombron and Jacquinot)." *Falkland Islands Dependencies Survey, Sci. Repts.*, no. 17.

[39] Spurr, E. B. 1974. "Individual Differences in Aggressiveness Among Adélie Penguins." *Anim. Beh.* 22:611–616.

[40] Spurr, E. B. 1975. "Breeding of the Adélie Penguin *Pygoscelis adeliae* at Cape Bird." *Ibis* 117:324–338.

[41] Spurr, E. B. 1975. "Orientation of Adélie Penguins on Their Territories." *Condor* 77:335–337.

[42]Spurr, E. B. 1975. "Communication in the Adélie Penguin." In B. Stonehouse, ed., *The Biology of Penguins,* pp. 449–501. London: Macmillan.

[43]Stonehouse, B. 1963. "Observations on Adélie Penguins (*Pygoscelis adeliae*) at Cape Royds, Antarctica." *Proc. Internatl. Ornithol. Congr.* 13:766–779.

[44]Taylor, R. H. 1962. "The Adélie Penguin *Pygoscelis adeliae* at Cape Royds." *Ibis* 104:176–204.

[45]Trillmich, F. 1978. "Feeding Territories and Breeding Success of South Polar Skuas." *Auk* 95:23–33.

[46]Trivelpiece, W. Z., S. G. Trivelpiece, G. R. Geupel, J. Kjelmyr, and N. J. Volkman. 1990. "Adélie and Chinstrap Penguins: Their Potential as Monitors of the Southern Ocean Marine Ecosystem." In K. R. Kerry and G. Hempel, eds., *Antarctic Ecosystems: Ecological Change and Conservation,* pp. 191–202. Berlin: Springer-Verlag.

[47]Trivelpiece, W. Z., S. G. Trivelpiece, and N. J. Volkman. 1987. "Ecological Segregation of Adélie, Gentoo, and Chinstrap Penguins at King George Island, Antarctica." *Ecology* 68:351–361.

[48]Volkman, N. J. and W. Trivelpiece. 1981. "Nest-Site Selection Among Adélie, Chinstrap and Gentoo Penguins in Mixed Species Rookeries." *Wilson Bull.* 93:243–248.

[49]Watanuki, Y. 1993. "Mortality of Eggs and Nest Attendance Pattern in Adélie Penguins in Lützow-Holm Bay." *Jap. J. Ornithol.* 42:1–8.

[50]White, M. G. and J. W. H. Conroy. 1975. "Aspects of Competition Between Pygoscelid Penguins at Signy Island, South Orkney Islands." *Ibis* 117:371–373.

[51]Whitehead, M. D., G. W. Johnstone, and H. R. Burton. 1990. "Annual Fluctuations in Productivity and Breeding Success of Adélie Penguins and Fulmarine Petrels in Prydz Bay, East Antarctica." In K. R. Kerry and G. Hempel, eds., *Antarctic Ecosystems: Ecological Change and Conservation,* pp. 214–223. Berlin: Springer-Verlag.

[52]Wood, R. C. 1971. "Population Dynamics of Breeding South Polar Skuas of Unknown Age." *Auk* 88:805–814.

[53]Wooller, R. D. and J. C. Coulson. 1977. "Factors Affecting the Age of First Breeding in the Kittiwake, *Rissa tridactyla*." *Ibis* 119:339–349.

[54]Yeates, G. W. 1968. "Studies on the Adélie Penguin at Cape Royds 1964–65 and 1965–66." *N.Z. J. Mar. Freshwater Res.* 2:472–496.

[55]Yeates, G. W. 1975. "Microclimate, Climate and Breeding Success in Antarctic Penguins." In B. Stonehouse, ed., *The Biology of Penguins,* pp. 397–409. London: Macmillan.

[56]Young, E. C. 1963. "Feeding Habits of the South Polar Skua *Catharacta maccormicki*." *Ibis* 105:301–318.

Chapter 6: The Reoccupation Period: Chicks and Breeding Success

[1]Ainley, D. G. 1975. "The Development of Reproductive Maturity in Adélie Penguins." In B. Stonehouse, ed., *The Biology of Penguins,* pp. 139–157. London: Macmillan.

[2]Ainley, D. G. 1975. "Displays of Adélie Penguins: A Reinterpretation." In B. Stonehouse, ed., *The Biology of Penguins,* pp. 503–534. London: Macmillan.

[3] Ainley, D. G. 1978. "Activity of Non-Breeding Adélie Penguins." *Condor* 80:135–146.

[4] Ainley, D. G. and D. P. DeMaster. 1980. "Survival and Mortality in a Population of Adélie Penguins." *Ecology* 61:522–530.

[5] Ainley, D. G., L. E. LeResche, and W. J. L. Sladen. 1983. *Breeding Biology of the Adélie Penguin.* Berkeley: University of California Press.

[6] Ainley, D. G., C. A. Ribic, and R. C. Wood. 1990. "A Demographic Study of the South Polar Skua at Cape Crozier." *J. Anim. Ecol.* 59:1–20.

[7] Ainley, D. G. and R. P. Schlatter. 1972. "Chick Raising Ability in Adélie Penguins." *Auk* 89:559–566.

[8] Ainley, D. G., P. R. Wilson, K. R. Barton, G. Ballard, N. Nur, and B. J. Karl. 1998. "Diet and Foraging Effort of Adélie Penguins in Relation to Pack-Ice Conditions in the Southern Ross Sea." *Polar Biol.* 20:311–319.

[9] Boekelheide, R. J. and D. G. Ainley. 1989. "Age, Resource Availability, and Breeding Effort in Brandt's Cormorant." *Auk* 106:389–401.

[10] Carrick, R. 1972. "Population Ecology of the Australian Magpie, Royal Penguin, and Silver Gull." In Population Ecology of Migratory Birds. *U.S. Dept. Interior, Wildl. Res. Rept.* 2:41–99.

[11] Coulson, J. C. 1963. "Egg Size and Shape in the Kittiwake (*Rissa tridactyla*) and Their Use in Estimating Age Composition of Populations." *Zool. Soc. London, Proc.* 140:211–227.

[12] Coulson, J. C. 1966. "The Influence of the Pair Bond and Age on the Breeding Biology of the Kittiwake Gull, *Rissa tridactyla*." *J. Anim. Ecol.* 35:269–279.

[13] Coulson, J. C. 1968. "Differences in the Quality of Birds Nesting in the Centre and the Edges of a Colony." *Nature* 217:478–479.

[14] Coulson, J. C. 1972. "The Significance of the Pair Bond in the Kittiwake." *Proc. 15th Internatl. Ornithol. Congr.* 424–433.

[15] Coulson, J. C. and J. M. Horobin. 1976. "The Influence of Age on the Breeding Biology and Survival of the Arctic Tern *Sterna paradisaea*." *J. Zool.* 178:247–260.

[16] Coulson, J. C. and E. White. 1958. "The Effect of Age on the Breeding Biology of the Kittiwake, *Rissa tridactyla*." *Ibis* 100:40–51.

[17] Croxall, J. P. and L. S. Davis. 1999. "Penguins: Paradoxes and Patterns." *Mar. Ornithol.* 27:1–12.

[18] Croxall, J. P., T. S. McCann, P. A. Prince, and P. Rothery. 1988. "Reproductive Performance of Seabirds and Seals at South Georgia and Signy Island, South Orkney Islands, 1976–1987: Implications for Southern Ocean Monitoring Studies." In D. Sahrhage, ed., *Antarctic Ocean and Resources Variability*, pp. 261–285. Berlin: Springer-Verlag.

[19] Cullen, J. M. 2000. "Divorce in Little Penguins." *Abstracts, Fourth Internatl. Penguin Conference*, p. 18. Coquimbo, Chile.

[20] Dann, P. and J. M. Cullen. 1990. "Survival, Patterns of Reproduction, and Lifetime Reproductive Output in Little Blue Penguins (*Eudyptula minor*) on Phillip Island, Victoria, Australia." In L. S. Davis and J. T. Darby, eds., *Penguin Biology*, pp. 63–84. Orlando, FL: Academic Press.

[21]Davis, L. S. 1982. "Creching Behaviour of Adélie Penguin Chicks (*Pygoscelis adeliae*)." *N.Z. J. Zool.* 9:279–286.

[22]Davis, L. S. 1982. "Timing of Nest Relief and Its Effect on Breeding Success in Adélie Penguins (*Pygoscelis adeliae*)." *Condor* 84:178–183.

[23]Davis, L. S. and F. T. McCaffrey. 1986. "Survival Analysis of Eggs and Chicks of Adélie Penguins (*Pygoscelis adeliae*)." *Auk* 103:379–388.

[24]Despin, B. 1977. "Croissances Comparées des Poussins chez les Manchots du Genre *Pygoscelis*." *C.R. Acad. Sci. Paris* 285, Ser. D:1135–1137.

[25]Goldsmith, R. 1962. "Reproductive Behaviour and Adaptation in the Adélie Penguin *Pygoscelis adeliae*." *J. Repro. Fert.* 4:237–238.

[26]Goldsmith, R. and W. J. L. Sladen. 1961. "Temperature Regulation of Some Antarctic Penguins." *J. Physiol.* 157:251–262.

[27]Irvine, L. G., J. R. Clarke, and K. R. Kerry. 2000. "Poor Breeding Success of the Adélie Penguin at Béchervaise Island in the 1998–99 Season." *CCAMLR Science* 7:151–167.

[28]Jablonski, B. 1985. "The Diet of Penguins on King George Island, South Shetland Islands." *Acta Zool. Cracov* 29:117–186.

[29]Kerry, K. R., J. R. Clarke, and G. D. Else. 1995. "The Foraging Range of Adélie Penguins at Béchervaise Island, MacRobertson Land, Antarctica, as Determined by Satellite Telemetry." In P. Dann, I. Norman, and P. Reilly, eds., *The Penguins: Ecology and Management,* pp. 216–143. Chipping North, NSW: Surrey Beatty.

[30]LeResche, R. E. and W. J. L. Sladen. 1970. "Establishment of Pair and Breeding Site Bonds by Known-Age Adélie Penguins." *Anim. Beh.* 18:517–526.

[31]Levick, G. M. 1914. *Antarctic Penguins: A Study of Their Social Habits.* New York: McBride, Nast.

[32]Lishman, G. S. 1985. "The Comparative Breeding Biology of Adélie and Chinstrap Penguins *Pygoscelis adeliae* and *P. antarctica* at Signy Island, South Orkney Islands." *Ibis* 127:84–99.

[33]Mills, J. A. 1973. "The Influence of Age and the Pair-Bond on Breeding of the Red-Billed Gull, *Larus novaehollandiae scolopinus*." *J. Anim. Ecol.* 42:147–162.

[34]Mills, J. A. 1979. "Factors Affecting the Egg Size of Red-Billed Gulls, *Larus novaehollandiae scolopinus*." *Ibis* 121:53–67.

[35]Nelson, J. B. 1966. "The Breeding Biology of the Gannet, *Sula bassana*, on the Bass Rock, Scotland." *Ibis* 108:584–626.

[36]Nelson, J. B. 1978. *The Gannet.* Vermilion, SD: Buteo Books.

[37]Nelson, J. B. 1978. *The Sulidae: Gannets and Boobies.* Oxford, U.K.: Oxford University Press.

[38]Ollason, J. C. and G. M. Dunnet. 1978. "Age, Experience and Other Factors Affecting the Breeding Success of the Fulmar, *Fulmarus glacialis*, in Orkney." *J. Anim. Ecol.* 47:961–976.

[39]Patterson, I. J. 1965. "Timing and Spacing of Broods in the Black-Headed Gull, *Larus ridibundus*." *Ibis* 107:433–459.

[40]Penney, R. L. 1968. "Territorial and Social Behavior in the Adélie Penguin." In O. A.

Austin, Jr., ed., Antarctic Bird Studies. *Antarc. Res. Ser.* 12:83–132. Washington, D.C.: American Geophysics Union.

[41] Pyle, P., L. B. Spear, W. J. Sydeman, and D. G. Ainley. 1991. "The Effects of Experience and Age on the Breeding Performance of Western Gulls." *Auk* 108:25–33.

[42] Reid, B. 1960. "New Zealander Studies Bird Life at Cape Hallett." *Antarctic* 2:211–213.

[43] Reid, B. 1965. "The Adélie Penguin (*Pygoscelis adeliae*) Egg." *N.Z. J. Sci.* 8:503–514.

[44] Reid, B. E. and C. Bailey. 1966. "The Value of the Yolk Reserve in Adélie Penguin Chicks." *Rec. Dominion Mus., Wellington* 5, no. 19:185–193.

[45] Richdale, L. E. 1957. *A Population Study of Penguins.* Oxford, U.K.: Oxford University Press.

[46] Sapin-Jaloustre, J. and F. Bourliere. 1951. "Incubation et Developpement du Poussin chez le Manchot Adélie *Pygoscelis adeliae.*" *Alauda* 19:65–83.

[47] Serventy, D. L. and P. J. Curry. 1984. "Observations on Colony Size, Breeding Success, Recruitment and Inter-Colony Dispersal in a Tasmanian Colony of Short-Tailed Shearwaters *Puffinus tenuirostris* Over a 30-Year Period." *Emu* 84:71–79.

[48] Shackleton, E. 1909. *The Heart of the Antarctic, the Farthest South Expedition.* London: William Heinemann.

[49] Sladen, W. J. L. 1958. "The Pygoscelid Penguins. I. Methods of Study. II. The Adélie Penguin *Pygoscelis adeliae* (Hombron and Jacquinot)." *Falkland Islands Dependencies Survey, Sci. Repts.,* no. 17.

[50] Spurr, E. B. 1975. "Breeding of the Adélie Penguin *Pygoscelis adeliae* at Cape Bird." *Ibis* 117:324–338.

[51] Spurr, E. B. 1975. "Orientation of Adélie Penguins on Their Territories." *Condor* 77:335–337.

[52] Spurr, E. B. 1975. "Behavior of the Adélie Penguin Chick." *Condor* 77:272–280.

[53] Spurr, E. B. 1977. "Adaptive Significance of the Reoccupation Period of the Adélie Penguin." In G. A. Llano, ed., *Adaptations Within Antarctic Ecosystems,* pp. 605–618. Houston: Gulf Publishing.

[54] Stonehouse, B. 1963. "Observations on Adélie Penguins (*Pygoscelis adeliae*) at Cape Royds, Antarctica." *Proc. Internatl. Ornithol. Congr.* 13:766–779.

[55] Sydeman, W. J., J. F. Penniman, T. M. Penniman, P. Pyle, and D. G. Ainley. 1991. "Breeding Performance of the Western Gull: Effects of Parental Age, Timing of Breeding, and Year in Relation to Food Availability." *J. Anim. Ecol.* 60:135–149.

[56] Taylor, R. H. 1962. "The Adélie Penguin *Pygoscelis adeliae* at Cape Royds." *Ibis* 104:176–204.

[57] Taylor, R. H. and H. S. Roberts. 1962. "Growth of Adélie Penguin (*Pygoscelis adeliae* Hombron and Jacquinot) Chicks." *N.Z. J. Sci.* 5:191–197.

[58] Tenaza, R. 1971. "Behavior and Nesting Success Relative to Nest Locations in Adélie Penguins (*Pygoscelis adeliae*)." *Condor* 73:81–92.

[59] Thompson, D. H. and J. T. Emlen. 1968. "Parent-Chick Individual Recognition in the Adélie Penguin." *Antarc. J. U.S.* 3:132.

[60]Trivelpiece, W. Z., S. G. Trivelpiece, G. R. Geupel, J. Kjelmyr, and N. J. Volkman. 1990. "Adélie and Chinstrap Penguins: Their Potential as Monitors of the Southern Ocean Marine Ecosystem." In K. R. Kerry and G. Hempel, eds., *Antarctic Ecosystems: Ecological Change and Conservation*, pp. 191–202. Berlin: Springer-Verlag.

[61]Trivelpiece, W. Z., S. G. Trivelpiece, and N. J. Volkman. 1987. "Ecological Segregation of Adélie, Gentoo, and Chinstrap Penguins at King George Island, Antarctica." *Ecology* 68:351–361.

[62]Volkman, N. J., P. Presler, and W. Trivelpiece. 1980. "Diets of Pygoscelid Penguins at King George Island, Antarctica." *Condor* 82:373–378.

[63]Volkman, N. J. and W. Trivelpiece. 1980. "Growth in the Pygoscelid Penguin Chicks." *J. Zool. (London)* 191:521–530.

[64]Volkman, N. J. and W. Trivelpiece. 1981. "Nest-Site Selection Among Adélie, Chinstrap and Gentoo Penguins in Mixed Species Rookeries." *Wilson Bull.* 93:243–248.

[65]Watanuki, Y. 1993. "Mortality of Eggs and Nest Attendance Pattern in Adélie Penguins in Lützow-Holm Bay." *Jap. J. Ornithol.* 42:1–8.

[66]Weimerskirch, H., J. Clobert, and P. Jouventin. 1987. "Survival in Five Southern Albatrosses and Its Relationship to Life History." *J. Anim. Ecol.* 56:1043–1055.

[67]Whitehead, M. D., G. W. Johnstone, and H. R. Burton. 1990. "Annual Fluctuations in Productivity and Breeding Success of Adélie Penguins and Fulmarine Petrels in Prydz Bay, East Antarctica." In K. R. Kerry and G. Hempel, eds., *Antarctic Ecosystems: Ecological Change and Conservation*, pp. 214–223. Berlin: Springer-Verlag.

[68]Williams, A. J. 1980. "Aspects of the Breeding Biology of the Gentoo Penguin *Pygoscelis papua*." *Le Gerfaut* 70:283–295.

[69]Williams, T. D. 1990. "Annual Variation in Breeding Biology of Gentoo Penguins, *Pygoscelis papua*, at Bird Island, South Georgia." *J. Zool. (London)* 222:247–258.

[70]Williams, T. D. and J. P. Croxall. 1991. "Chick Growth and Survival in Gentoo Penguins (*Pygoscelis papua*): Effect of Hatching Asynchrony and Variation in Food Supply." *Polar Biol.* 11:197–202.

[71]Wood, R. C. 1971. "Population Dynamics of Breeding South Polar Skuas of Unknown Age." *Auk* 88:805–814.

[72]Wooller, R. D. and J. C. Coulson. 1977. "Factors Affecting the Age of First Breeding in the Kittiwake, *Rissa tridactyla*." *Ibis* 119:339–349.

[73]Yeates, G. W. 1968. "Studies on the Adélie Penguin at Cape Royds 1964–65 and 1965–66." *N.Z. J. Mar. Freshwater Res.* 2:472–496.

[74]Young, E. C. 1994. *Skua and Penguin: Predator and Prey*. Cambridge, U.K.: Cambridge University Press.

Chapter 7: Predation

[1]Ainley, D. G. 1974. "The Comfort Behaviour of Adélie and Other Penguins." *Behaviour* (Ser. L) 1–2:16–51.

[2]Ainley, D. G., S. H. Morrell, and R. C. Wood. 1986. "South Polar Skua Breeding Colonies in the Ross Sea Region, Antarctica." *Notornis* 33:155–163.

[3]Ainley, D. G., E. F. O'Connor, and R. J. Boekelheide. 1984. "The Marine Ecology of Birds in the Ross Sea, Antarctica." *Amer. Ornithol. Union, Monogr.* no. 32.

[4]Ainley, D. G., L. B. Spear, and R. C. Wood. 1985. "Sexual Color and Size Dimorphism in the South Polar Skua." *Condor* 87:427–428.

[5]Condy, P. R., R. J. van Aarde, and M. N. Bester. 1978. "The Seasonal Occurrence and Behaviour of Killer Whales *Orcinus orca* at Marion Island." *J. Zool., Lond.* 184:449–464.

[6]Court, G. S. 1996. "The Seal's Own Skin Game." *Nat. Hist.* 105:36–41.

[7]Croxall, J. P., P. A. Prince, I. Hunter, S. J. McInnes, and P. G. Copestake. 1984. "The Seabirds of the Antarctic Peninsula, Islands of the Scotia Sea, and Antarctic Continent Between 80°W and 20°W: Their Status and Conservation." In J. P. Croxall, P. G. H. Evans, and R. W. Schreiber, eds., *Status and Conservation of the World's Seabirds,* pp. 637–666, Tech. Publ. no. 2. Cambridge, U.K.: International Council of Bird Preservation.

[8]Eklund, C. R. 1961. "The Antarctic Skua." *Nat. Hist.* 70:92–100.

[9]Guinet, C. 1992. "Comportement de Chasse des Orques (*Orcinus orca*) Autour des Iles Crozet." *Canad. J. Zool.* 70:1656–1667.

[10]Guinet, C. and J. Bouvier. 1995. "Development of Intentional Stranding Hunting Techniques in Killer Whale (*Orcinus orca*) Calves at Crozet Archipelago." *Canad. J. Zool.* 73:27–33.

[11]Hemings, A. D. 1984. "Aspects of the Breeding Biology of McCormick's Skua *Catharacta maccormicki* at Signy Island, South Orkney Islands." *Brit. Antarc. Surv. Bull.* 65:65–79.

[12]Hofmann, R. J., R. A. Reichle, D. B. Siniff, and D. Muller-Schwarze. 1977. "The Leopard Seal (*Hydrurga leptonyx*) at Palmer Station, Antarctica." In G. A. Llano, ed., *Adaptations Within Antarctic Ecosystems,* pp. 769–782. Washington, D.C.: Smithsonian Institution.

[13]Kooyman, G. L. 1965. "Leopard Seals of Cape Crozier." *Animals* 6:59–63.

[14]Kooyman, G. L. 1981. "Leopard Seal *Hydrurga leptonyx* Blainville, 1820." In S. H. Ridgway and R. J. Harrison, eds., *Handbook of Marine Mammals,* pp. 261–274. London: Academic Press.

[15]Levick, G. M. 1914. *Antarctic Penguins: A Study of Their Social Habits.* New York: McBride, Nast.

[16]Lowe, P. R. and N. B. Kinnear. 1930. "Birds." *British Antarc. ("Terra Nova") Exped., 1910, Nat. Hist. Rept., Zool.,* 4, no. 5:103–193. London: British Museum.

[17]Maher, W. J. 1966. "Predation Impact on Penguins." *Nat. Hist.* 75:42–51.

[18]Mikhalev, Y. A., M. V. Ivahin, V. P. Savusin, and F. E. Zelenaya. 1981. "The Distribution and Biology of Killer Whales in the Southern Hemisphere." *Rept. Internatl. Whal. Commn.* 31:551–566.

[19]Muller-Schwarze, D. and C. Muller-Schwarze. 1975. "Relations Between Leopard Seals and Adélie Penguins." *Rapp. P.-v. Reun. Cons. Int. Explor. Mer.* 169:394–404.

[20]Muller-Schwarze, D. and C. Muller-Schwarze. 1977. "Interactions Between South

Polar Skuas and Adélie Penguins." In G. A. Llano, ed., *Adaptations Within Antarctic Ecosystems*, pp. 619–647. Washington, D.C.: Smithsonian Institution.

[21] Murphy, R. C. 1936. *The Oceanic Birds of South America.* New York: Macmillan.

[22] Oritsland, T. 1977. "Food Consumption of Seals in the Antarctic Pack Ice." In G. A. Llano, ed., *Adaptations Within Antarctic Ecosystems,* pp. 749–768. Washington, D.C.: Smithsonian Institution.

[23] Penney, R. L. and G. Lowry. 1967. "Leopard Seal Predation on Adélie Penguins." *Ecology* 48:878–882.

[24] Pietz, P. J. 1987. "Feeding and Nesting Ecology of Sympatric South Polar and Brown Skuas." *Auk* 104:617–627.

[25] Ponting, H. G. 1921. *The Great White South: Or with Scott in the Antarctic.* London: Duckworth.

[26] Prévost, J. 1961. *Ecologie du Manchot Empereur* Aptenodytes forsteri *Gray.* Paris: Hermann.

[27] Randall, R. M. and B. M. Randall. 1990. "Cetaceans as Predators of Jackass Penguins Spheniscus demersus: Deductions Based on Behaviour." *Mar. Ornithol.* 18:9–12.

[28] Rogers, T. and M. M. Bryden. 1995. "Predation of Adélie Penguins (*Pygoscelis adeliae*) by Leopard Seals (*Hydrurga leptonyx*) in Prydz Bay, Antarctica." *Canad. J. Zool.* 73:1001–1004.

[29] Rounsevell, D. E. and G. R. Copson. 1982. "Growth Rate and Recovery of a King Penguin, *Aptenodytes patagonicus,* Population After Exploitation." *Austr. Wildl. Res.* 9:519–525.

[30] Sladen, W. J. L. 1958. "The Pygoscelid Penguins. I. Methods of Study. II. The Adélie Penguin *Pygoscelis adeliae* (Hombron and Jacquinot)." *Falkland Islands Dependencies Survey, Sci. Repts.,* no. 17.

[31] Spurr, E. B. 1975. "Orientation of Adélie Penguins on Their Territories." *Condor* 77:335–337.

[32] Thomas, J. A., S. Leatherwood, W. E. Evans, J. R. Jehl, Jr., and F. T. Awbry. 1981. "Ross Sea Killer Whale Distribution, Behavior, Color Patterns, and Vocalizations." *Antarc. J. U.S. (Annual Review)* 16:57–158.

[33] Trillmich, F. 1978. "Feeding Territories and Breeding Success of South Polar Skuas." *Auk* 95:23–33.

[34] Trivelpiece, W. and N. J. Volkman. 1982. "Feeding Strategies of Sympatric South Polar *Catharacta maccormicki* and Brown Skuas *Catharacta lönnbergi.*" *Ibis* 124:50–54.

[35] Walker, E. D. 1968. *Mammals of the World,* 2nd ed. Baltimore, MD: Johns Hopkins University Press.

[36] Williams, A. J., B. M. Dyer, R. M. Randall, and J. Komen. 1990. "Killer Whales *Orcinus orca* and Seabirds: 'Play,' Predation and Association." *Mar. Ornithol.* 18:37–41.

[37] Wilson, E. A. 1907. "Aves." *Natl. Antarctic Exped. 1901–04, Vol. 2. Zoology,* Part 2:1–121. London: British Museum.

[38] Wilson, P. R., D. G. Ainley, N. Nur, S. S. Jacobs, K. J. Barton, G. Ballard, and J. C. Comiso. 2001. "Adélie Penguin Population Change in the Pacific Sector of Antarc-

tica: Relation to Sea-Ice Extent and the Antarctic Circumpolar Current." *Mar. Ecol. Progr. Ser.* 213:301–309.

[39] Yeates, G. W. 1971. "Observations on Orientation of Penguins to Wind and on Colonization in the Adélie Penguin Rookery at Cape Royds, Antarctica." *N.Z. J. Sci.* 14:901–906.

[40] Young, E. C. 1963. "Feeding Habits of the South Polar Skua *Catharacta maccormicki*." *Ibis* 105:301–318.

[41] Young, E. C. 1994. *Skua and Penguin: Predator and Prey*. Cambridge, U.K.: Cambridge University Press.

Chapter 8: Demography

[1] Ainley, D. G. 1978. "Activity of Non-Breeding Adélie Penguins." *Condor* 80:135–146.

[2] Ainley, D. G. 1980. "Seabirds as Marine Organisms: A Review." *Calif. Co-op. Ocean. Fish. Investig., Rept.* 23:48–53.

[3] Ainley, D. G. and D. P. DeMaster. 1980. "Survival and Mortality in a Population of Adélie Penguins." *Ecology* 61:522–530.

[4] Ainley, D. G., L. E. LeResche, and W. J. L. Sladen. 1983. *Breeding Biology of the Adélie Penguin*. Berkeley: University of California Press.

[5] Ainley, D. G., E. F. O'Connor, and R. J. Boekelheide. 1984. "The Marine Ecology of Birds in the Ross Sea, Antarctica." *Amer. Ornithol. Union,* Monogr. no. 32.

[6] Ainley, D. G., R. Podolsky, L. DeForest, G. Spencer, and N. Nur. 2001. "The Status and Population Trends of the Newell's Shearwater on Kauai: Insights from Modeling." In M. Scott, ed., Conservation of Hawaiian Birds. *Stud. Avian Biol.* 22:108–123.

[7] Ainley, D. G., C. A. Ribic, and W. R. Fraser. 1992. "Does Prey Preference Affect Habitat Choice in Antarctic Seabirds?" *Mar. Ecol. Progr. Ser.* 90:207–221.

[8] Ainley, D. G., W. J. Sydeman, and J. Norton. 1995. "Upper-Trophic Level Predators Indicate Interannual Negative and Positive Anomalies in the California Current Food Web." *Mar. Ecol. Progr. Ser.* 118:69–79.

[9] Ashmole, N. P. 1963. "The Regulation of Numbers of Tropical Oceanic Birds." *Ibis* 103b:458–473.

[10] Blackburn, N., R. H. Taylor, and P. R. Wilson. 1991. "An Interpretation of the Growth of the Adélie Penguin Rookery at Cape Royds, 1955–1990." *N.Z. J. Ecol.* 15, no. 2:23–28.

[11] Boekelheide, R. J. and D. G. Ainley. 1989. "Age, Resource Availability, and Breeding Effort in Brandt's Cormorant." *Auk* 106:389–401.

[12] Brown, J. L. 1969. "Territorial Behavior and Population Regulation in Birds." *Wilson Bull.* 81:293–329.

[13] Carrick, R. 1972. "Population Ecology of the Australian Magpie, Royal Penguin, and Silver Gull." In Population Ecology of Migratory Birds. *U.S. Dept. Interior, Wildl. Res. Rept.* 2:41–99.

[14] Coulson, J. C. 1991. "The Population Dynamics of Culling Herring Gulls and Lesser Black-Backed Gulls." In C. M. Perrins, J.-D. Lebreton, and G. J. M. Hirons, eds.,

  Bird Population Studies: The Relevance to Conservation and Management, pp. 479–496. Oxford, U.K.: Oxford University Press.

[15]Coulson, J. C. and R. D. Wooler. 1976. "Differential Survival Rates Among Breeding Kittiwake Gulls." *J. Anim. Ecol.* 45:205–213.

[16]Crawford, R. J. M., J. H. M. David, A. J. Williams, and B. M. Dyer. 1989. "Competition for Space: Recolonising Seals Displace Endangered, Endemic Seabirds Off Namibia." *Biolog. Conserv.* 48:59–72.

[17]Crawford, R. J. M., L. J. Shannon, and P. A. Whittington. 1999. "Population Dynamics of the African Penguin *Spheniscus demersus* at Robben Island, South Africa." *Mar. Ornithol.* 27:139–147.

[18]Crawford, R. J. M., A. J. Williams, R. M. Randall, B. M. Randall, A. Berruti, and G. J. B. Ross. 1990. "Recent Population Trends of Jackass Penguins *Spheniscus demersus* off Southern Africa." *Biolog. Conserv.* 52:229–243.

[19]Croxall, J. P. and L. S. Davis. 1999. "Penguins: Paradoxes and Patterns." *Mar. Ornithol.* 27:1–12.

[20]Dann, P. and J. M. Cullen. 1990. "Survival, Patterns of Reproduction, and Lifetime Reproductive Output in Little Blue Penguins (*Eudyptula minor*) on Phillip Island, Victoria, Australia." In L. S. Davis and J. T. Darby, eds., *Penguin Biology,* pp. 63–84. Orlando, FL: Academic Press.

[21]Darby, J. T. and P. J. Seddon. 1990. "Breeding Biology of Yellow-Eyed Penguins (*Megadyptes antipodes*)." In L. S. Davis and J. T. Darby, eds., *Penguin Biology,* pp. 45–62. Orlando, FL: Academic Press.

[22]Dunnet, G. M. and J. C. Ollason. 1978. "The Estimation of Survival Rates in the Fulmar, *Fulmarus glacialis.*" *J. Anim. Ecol.* 47:507–520.

[23]Furness, R. W. and T. R. Birkhead. 1984. "Seabird Colony Distributions Suggest Competition for Food Supplies During the Breeding Season." *Nature* 311:655–656.

[24]Gaston, A. J., G. Chapelaine, and D. G. Noble. 1983. "The Growth of Thick-Billed Murre Chicks at Colonies in Hudson Strait: Inter- and Intra-Colony Variation." *Canad. J. Zool.* 61:2465–2475.

[25]Goodman, D. 1974. "Natural Selection and a Cost Ceiling on Reproductive Effort." *Amer. Natural.* 108:247–268.

[26]Hofman, R. J., R. A. Reichle, D. B. Siniff, and D. Muller-Schwarze. 1977. "The Leopard Seal (*Hydrurga leptonyx*) at Palmer Station, Antarctica." In G. A. Llano, ed., *Adaptations Within Antarctic Ecosystems,* pp. 769–782. Houston, TX: Gulf Publishing.

[27]Hunt, G. L., Jr., Z. A. Eppley, and D. C. Schneider. 1986. "Reproductive Performance of Seabirds: The Importance of Population and Colony Size." *Auk* 103:306–317.

[28]Kadlec, J. A. and W. H. Drury. 1968. "Structure of the New England Herring Gull Population." *Ecology* 49:644–676.

[29]King, J. E. 1964. *Seals of the World.* London: British Museum of Natural History.

[30]Lack, D. 1954. *The Natural Regulation of Animal Numbers.* Oxford, U.K.: Clarendon Press.

[31] Lack, D. 1966. *Population Studies of Birds.* Oxford, U.K.: Clarendon Press.

[32] Lack, D. 1968. *Adaptations for Breeding in Birds.* London: Methuen.

[33] Leslie, P. H. 1949. "On the Use of Certain Matrices in Population Mathematics." *Biometrika* 33:183–212.

[34] Linden, M. and A. P. Møller. 1989. "Cost of Reproduction and Covariation of Life History Traits in Birds." *Trends Ecol. Evol.* 4:367–371.

[35] Manuwal, D. 1974. "Effects of Territoriality on Breeding in a Population of Cassin's Auklet." *Ecology* 55:1399–1406.

[36] Marr, J. W. S. 1962. "The Natural History and Geography of the Antarctic Krill (*Euphausia superba* Dana)." *Discovery Rept.* 32:33–464.

[37] Mills, J. A. 1989. "Red-Billed Gull." In I. Newton, ed., *Lifetime Reproduction in Birds,* pp. 387–404. New York: Academic Press.

[38] Murphy, R. C. 1925. *The Bird Islands of Peru.* New York: Putnam.

[39] Newton, I. 1989. "Synthesis." In I. Newton, ed., *Lifetime Reproduction in Birds,* pp. 441–469. New York: Academic Press.

[40] Newton, I. 1991. "Concluding Remarks." In C. M. Perrins, J.-D. Lebreton, and G. J. M. Hirons, eds., *Bird Population Studies: The Relevance to Conservation and Management,* pp. 637–654. Oxford, U.K.: Oxford University Press.

[41] Ollason, J. A. and G. M. Dunnet. 1978. "Age, Experience and Other Factors Affecting the Breeding Success of the Fulmar, *Fulmarus glacialis* in Orkney." *J. Anim. Ecol.* 47:961–976.

[42] Partridge, L. 1989. "Lifetime Reproductive Success and Life-History Evolution." In I. Newton, ed., *Lifetime Reproduction in Birds,* pp. 421–440. New York: Academic Press.

[43] Perrins, C. M. 1991. "Constraints on the Demographic Parameters of Bird Populations." In C. M. Perrins, J.-D. Lebreton, and G. J. M. Hirons, eds., *Bird Population Studies: The Relevance to Conservation and Management,* pp. 191–206. Oxford, U.K.: Oxford University Press.

[44] Randall, R. M. 1983. *Biology of the Jackass Penguin* Spheniscus demersus *(L.) at St. Croix Island, South Africa.* Unpublished Ph.D. dissertation, University of Port Elizabeth, South Africa.

[45] Reid, B. 1968. "An Interpretation of the Age Structure and Breeding Status of an Adélie Penguin Population." *Notornis* 15:193–197.

[46] Richdale, L. E. 1957. *A Population Study of Penguins.* Oxford, U.K.: Oxford University Press.

[47] Ricklefs, R. E. 1973. "Fecundity, Mortality and Avian Demography." In D. S. Farner, ed., *Breeding Biology of Birds,* pp. 366–435. Washington, D.C.: National Academy of Science.

[48] Ricklefs, R. E. 1983. "Comparative Avian Demography." In R. F. Johnston, ed., *Current Ornithology,* vol. 1. New York: Plenum.

[49] Shannon, L. J. and R. J. M. Crawford. 1999. "Management of the African Penguin *Spheniscus demersus:* Insights from Modeling." *Mar. Ornithol.* 27:119–128.

[50] Sladen, W. J. L. 1958. "The Pygoscelid Penguins. I. Methods of Study. II. The Adélie

Penguin *Pygoscelis adeliae* (Hombron and Jacquinot)." *Falkland Islands Dependencies Survey, Sci. Repts.*, no. 17.

[51]Sladen, W. J. L. and R. E. LeResche. 1970. "New and Developing Techniques in Antarctic Ornithology." In M. Holdgate, ed., *Antarctic Ecology*, vol. 1, pp. 585–604. New York: Academic Press.

[52]Spellerberg, I. F. 1975. "The Predators of Penguins." In B. Stonehouse, ed., *The Biology of Penguins*, pp. 413–434. London: Macmillan.

[53]Spurr, E. B. 1975. "Breeding of the Adélie Penguin *Pygoscelis adeliae* at Cape Bird." *Ibis* 117:324–338.

[54]Stirling, I. 1971. "Population Dynamics of the Weddell Seal (*Leptonychotes weddelli*) in McMurdo Sound, Antarctica." In W. H. Burt, ed., Antarctic Pinnipeds. *Antarc. Res. Ser.* 18:141–161. Washington, D.C.: American Geophysics Union.

[55]Stonehouse, B. 1960. "The King Penguin, *Aptenodytes patagonica*, of South Georgia, Pt. 1." *Falkland Islands Dependencies Surveys, Sci. Repts.*, no. 23.

[56]Stonehouse, B. 1967. "Occurrence and Effects of Open Water in McMurdo Sound, Antarctica, During Winter and Early Spring." *Polar Rec.* 13:775–778.

[57]von Haartman, L. 1971. "Population Dynamics." In D. S. Farner, J. R. King, and K. C. Parkes, eds., *Avian Biology*, vol. 1, pp. 391–459. New York: Academic Press.

[58]Whitehead, M. D., G. W. Johnstone, and H. R. Burton. 1990. "Annual Fluctuations in Productivity and Breeding Success of Adélie Penguins and Fulmarine Petrels in Prydz Bay, East Antarctica." In K. R. Kerry and G. Hempel, eds., *Antarctic Ecosystems: Ecological Change and Conservation*, pp. 214–223. Berlin: Springer-Verlag.

[59]Williams, G. C. 1966. "Natural Selection, the Cost of Reproduction and a Refinement of Lack's Principle." *Amer. Natural.* 100:687–690.

[60]Wilson, P. R., D. G. Ainley, N. Nur, S. S. Jacobs, K. J. Barton, G. Ballard, and J. C. Comiso. 2001. "Adélie Penguin Population Growth in the Pacific Sector of Antarctica: Relation to Sea-Ice Extent and the Southern Oscillation." *Mar. Ecol. Progr. Ser.* 213:301–309.

[61]Wooller, R. D., J. S. Bradley, I. J. Kira, and D. L. Serventy. 1989. "Short-Tailed Shearwater." In I. Newton, ed., *Lifetime Reproduction in Birds*, pp. 405–417. New York: Academic Press.

[62]Wooller, R. D. and J. C. Coulson. 1977. "Factors Affecting the Age of First Breeding in the Kittiwake, *Rissa tridactyla*." *Ibis* 119:339–349.

Chapter 9: The Bellwether of Climate Change

[1]Ackley, S. F., P. Wadhams, and J. Comiso. 2001. "Increases of Antarctic Sea Ice Extent from Whaling Records: Some Inconsistencies with the Historical and Modern Records." Abstract and Poster, Gordon Research Conference on Polar Marine Science, March 11–15, 2001, Ventura, Calif.

[2]Ainley, D. G., L. E. LeResche, and W. J. L. Sladen. 1983. *Breeding Biology of the Adélie Penguin*. Berkeley: University of California Press.

[3]Ainley, D. G., P. R. Wilson, K. J. Barton, G. Ballard, N. Nur, and B. Karl. 1998. "Diet

and Foraging Effort of Adélie Penguins in Relation to Pack-Ice Conditions in the Southern Ross Sea." *Polar Biol.* 20:311–319.

[4] Barbraud, C. and H. Weimerskirch. 2001. "Emperor Penguins and Climate Change." *Nature* 411:183–186.

[5] Baroni, C. and G. Orombelli. 1991. "Holocene Raised Beaches at Terra Nova Bay, Victoria Land, Antarctica." *Quaternary Res.* 36:157–177.

[6] Baroni, C. and G. Orombelli. 1994. "Abandoned Penguin Rookeries as Holocene Paleoclimatic Indicators in Antarctica." *Geology* 22:23–26.

[7] Berkman, P. A., J. T. Andrews, S. Bjorcke, E. A. Colhoun, S. D. Emslie, I. D. Goodwin, B. L. Hall, C. P. Hart, K. Hirakawa, A. Igarashi, O. Ingolfsson, J. Lopez-Martinez, W. B. Lyons, M. C. G. Mabin, P. G. Quilty, M. Taviani, and Y. Yoshida. 1998. "Circum-Antarctic Coastal Environmental Shifts During the Late Quaternary Reflected by Emerged Marine Deposits." *Antarc. Sci.* 10:345–362.

[8] Bonner, W. N. 1984. "Conservation and the Antarctic." In R. M. Laws, ed., *Antarctic Ecology*, vol. 2, pp. 821–850. London: Academic Press.

[9] Brown, S. G. and C. H. Lockyer. 1984. "Whales." In R. M. Laws, ed., *Antarctic Ecology*, vol. 2, pp. 717–781. London: Academic Press.

[10] Burkle, L. H. 2000. "Determining Sea Ice Distribution in the Southern Ocean During the Last Glacial Maximum." In Sea Ice and Its Interactions with the Ocean, Atmosphere, and Biosphere (Abstract). *Internatl. Glaciolog. Soc. Symp.*, Fairbanks, AK, June 19–23, 2000.

[11] Comiso, J. C. 2000. "Variability and Trends in Antarctic Surface Temperatures from in Situ and Satellite Infrared Measurements." *J. Climate* 13:1674–1696.

[12] Conway, H., B. L. Hall, G. H. Denton, A. M. Gades, and E. D. Waddington. 1999. "Past and Future Grounding-Line Retreat of the West Antarctic Ice Sheet." *Science* 286:280–283.

[13] Croxall, J. P. 1992. "Southern Ocean Environmental Changes: Effects on Seabird, Seal and Whale Populations." *Phil. Trans. Royal Soc. London* B338:319–328.

[14] Croxall, J. P., D. M. Rootes, and P. A. Prince. 1981. "Increases in Penguin Populations at Signy Island, South Orkney Islands." *Br. Antarc. Surv. Bull.* 54:47–56.

[15] de la Mare, W. K. 1997. "Abrupt Mid-Twentieth-Century Decline in Antarctic Sea-Ice Extent from Whaling Records." *Nature* 389:57–60.

[16] Doake, C. S. M. and D. G. Vaughan. 1991. "Rapid Disintegration of the Wordie Ice Shelf in Response to Atmospheric Warming." *Nature* 350:328–330.

[17] Emslie, S. D. 2001. "Assessing Radiocarbon Dates from Abandoned Penguin Colonies in the Antarctic Peninsula." *Antarc. Sci.* 13:289–295.

[18] Emslie, S. D., W. Fraser, R. C. Smith, and W. Walker. 1998. "Abandoned Penguin Colonies and Environmental Change in the Palmer Station Area, Anvers Island, Antarctic Peninsula." *Antarc. Sci.* 10:257–268.

[19] Emslie, S. D. and J. D. McDaniel. 2002. "Adélie Penguin Diet and Climate Change During the Middle to Late Holocene in Northern Marguerite Bay, Antarctic Peninsula." *Polar Biol.* 25:222–229.

[20] Folland, C. K., D. E. Parker, and F. E. Kates. 1984. "World-Wide Marine Temperature Fluctuations 1856–1981." *Nature* 310:670–673.

[21] Fraser, W. R. and D. L. Patterson. 1997. "Human Disturbance and Long-Term Changes in Adélie Penguin Populations: A Natural Experiment at Palmer Station, Antarctica." In B. Battaglia, J. Valencia, and D. W. H. Walton, eds., *Antarctic Communities: Species, Structure and Survival*, pp. 445–452. Cambridge, U.K.: Cambridge University Press.

[22] Fraser, W. R., W. Z. Trivelpiece, D. G. Ainley, and S. G. Trivelpiece. 1992. "Increases in Antarctic Penguin Populations: Reduced Competition with Whales or a Loss of Sea Ice Due to Environmental Warming?" *Polar Biol.* 11:525–531.

[23] Goodwin, I. D. 1993. "Holocene Deglaciation, Sea-Level Change, and Emergence of the Windmill Islands, Budd Coast, Antarctica." *Quaternary Res.* 40:70–80.

[24] Hansen, J. and S. Lebedeff. 1987. "Global Trends of Measured Surface Air Temperature." *J. Geophys. Res.* 92:13345–13372.

[25] Ingolfsson, O., C. Hjort, P. A. Berkman, S. Bjorck, E. Colhoun, I. D. Goodwin, B. Hall, K. Hirakawa, P. Melles, P. Moller, and M. I. Prentice. 1998. "Antarctic Glacial History Since the Last Glacial Maximum: An Overview of the Record on Land." *Antarc. Sci.* 10:326–344.

[26] Ingram, W. J., C. A. Wilson, and J. F. B. Mitchell. 1989. "Modeling Climate Change: An Assessment of Sea Ice and Surface Albedo Feedbacks." *J. Geophys. Res.* 94, no. D6:8609–8622.

[27] Jacobs, S. S. and J. C. Comiso. 1993. "A Recent Sea-Ice Retreat West of the Antarctic Peninsula." *Geophys. Res. Letters* 20, no. 12:1171–1174.

[28] Jacobs, S. S. and J. C. Comiso. 1997. "Climate Variability in the Amundsen and Bellingshausen Seas." *J. Climate* 10:697–709.

[29] Jones, P. D. 1990. "Antarctic Temperatures Over the Present Century (a Study of the Early Expedition Record)." *J. Climate* 3:1193–1203.

[30] Jones, P. D., S. C. B. Raper, and T. M. L. Wigley. 1986. "Southern Hemisphere Surface Air Temperature Variations: 1851–1984." *J. Climate Appl. Meteorol.* 25:1213–1230.

[31] Keys, H., S. S. Jacobs, and L. W. Brigham. 1998. "Continued Northward Expansion of the Ross Ice Shelf." *Annals Glaciol.* 27:93–98.

[32] Laws, R. M. 1977. "Seals and Whales of the Southern Ocean." *Phil. Trans Royal Soc.* B279:81–96.

[33] Laws, R. M. 1977. "The Significance of Vertebrates in the Antarctic Marine Ecosystem." In G. A. Llano, ed., *Adaptations Within Antarctic Ecosystems*, pp. 411–438. Houston, TX: Gulf Publishing.

[34] Manabe, S., R. J. Stouffer, M. J. Spelman, and K. Bryan. 1991. "Transient Responses of a Coupled Ocean-Atmosphere Model to Gradual Changes of Atmospheric CO_2. Part I: Annual Mean Response." *J. Climate* 4:785–818.

[35] Martin, R. R., G. W. Johnstone, and E. J. Woehler. 1990. "Increased Numbers of Adélie Penguins *Pygoscelis adeliae* Breeding Near Casey, Wilkes Land, East Antarctica." *Corella* 14:119–123.

[36]Micol, T. and P. Jouventin. 2001. "Long-Term Population Trends in Seven Antarctic Seabirds at Pointe Géologie (Terre Adélie)." *Polar Biol.* 24:647–656.

[37]Nicol, S., T. Pauly, N. L. Bindoff, S. Wright, D. Thiele, G. W. Hosie, P. G. Strutton, and E. Woehler. 2000. "Ocean Circulation Off East Antarctica Affects Ecosystem Structure and Sea-Ice Extent." *Nature* 406:504–507.

[38]Petit, J. R., J. Jouzel, D. Raynaud, N. I. Barkov, J.-M. Barnola, I. Basile, M. Benders, J. Chappellaz, M. Davis, G. Delaygue, M. Delmotte, V. M. Kotlyakov, M. Legrand, V. Y. Lipenkov, C. Lorius, L. Pépin, C. Ritz, E. Saltzman, and M. Stievenard. 1999. "Climate and Atmospheric History of the Past 420,000 Years from the Vostok Ice Core, Antarctica." *Nature* 399:429–436.

[39]Rind, D., R. Healy, C. Parkinson, and D. Martinson. 1995. "The Role of Sea Ice in 2 X CO_2 Climate Model Sensitivity. Part I: The Total Influence of Sea Ice Thickness and Extent." *J. Climate* 8:449–463.

[40]Roeder, A. D., R. K. Marshall, A. J. Mitchelson, T. Visagathlagar, P. A. Richie, D. R. Love, T. J. Pakai, H. C. McPartlan, N. D. Murray, N. A. Robinson, K. R. Kerry, and D. M. Lambert. 2001. "Gene Flow on the Ice: Genetic Differentiation Among Adélie Penguin Colonies Around Antarctica." *Molecular Ecology* 10:1645–1656.

[41]Simpson, G. G. 1974. "Fossil Penguins." In B. Stonehouse, ed., *The Biology of Penguins*, pp. 19–41. London: Macmillan.

[42]Sladen, W. J. L. 1958. "The Pygoscelid Penguins. I. Methods of Study. II. The Adélie Penguin *Pygoscelis adeliae* (Hombron and Jacquinot)." *Falkland Islands Dependencies Survey, Sci. Repts.*, no. 17.

[43]Smith, R. C., D. Ainley, K. Baker, E. Domack, S. Emslie, B. Fraser, J. Kennett, A. Leventer, E. Mosley-Thompson, S. Stammerjohn, and M. Vernet. 1999. "Marine Ecosystem Sensitivity to Climate Change." *BioScience* 49, no. 5:393–404.

[44]Speir, T. W. and J. C. Cowling. 1984. "Ornithologenic Soils of the Cape Bird Adélie Penguin Rookeries, Antarctica. I. Chemical Properties." *Polar Biol.* 2:199–205.

[45]Spellerberg, I. F. 1970. "Abandoned Penguin Rookeries Near Cape Royds, Ross Island, Antarctica and ^{14}C Dating of Penguin Remains." *N.Z. J. Science* 13:380–385.

[46]Stammerjohn, S. E. and R. C. Smith. 1997. "Opposing Southern Ocean Climate Patterns as Revealed by Trends in Regional Sea Ice Coverage." *Climate Change* 37:617–639.

[47]Stonehouse, B. 1970. "Recent Climatic Change in Antarctica Suggested from ^{14}C Dating of Penguin Remains." *Palaeogeogr. Palaeoclim. Palaeoecol.* 7:341–343.

[48]Stuiver, M., G. H. Denton, T. Hughes, and J. L. Fastook. 1981. "History of the Marine Ice Sheet in West Antarctica During the Last Glaciation: A Working Hypothesis." In G. H. Denton and T. Hughes, eds., *The Last Great Ice Sheets*, pp. 319–369. New York: Wiley.

[49]Tatur, A., A. Myrcha, and J. Niegodzisz. 1997. "Formation of Abandoned Penguin Rookery Ecosystems in the Maritime Antarctic." *Polar Biol.* 17:405–417.

[50]Taylor, R. H. and P. R. Wilson. 1990. "Recent Increase and Southern Expansion of

Adélie Penguin Populations in the Ross Sea, Antarctica, Related to Climatic Warming." *N.Z. J. Ecol.* 14:25–29.

[51] Taylor, R. H., P. R. Wilson, and B. W. Thomas. 1990. "Status and Trends of Adélie Penguin Populations in the Ross Sea Region." *Polar Rec.* 26, no. 159:293–304.

[52] Thomson, R. B. 1977. "Effects of Human Disturbance on an Adélie Penguin Rookery and Measures of Control." In G. A. Llano, ed., *Adaptations Within Antarctic Ecosystems*, pp. 1177–1180. Houston, TX: Gulf Publishing.

[53] Trathan, P. N., J. P. Croxall, and E. J. Murphy. 1996. "Dynamics of Antarctic Penguin Populations in Relation to Inter-Annual Variability in Sea Ice Distribution." *Polar Biol.* 16:321–330.

[54] Trivelpiece, W. Z., S. G. Trivelpiece, G. R. Geupel, J. Kjelmyr, and N. J. Volkman. 1990. "Adélie and Chinstrap Penguins: Their Potential as Monitors of the Southern Ocean Marine Ecosystem." In K. R. Kerry and G. Hempel, eds., *Antarctic Ecosystems: Ecological Change and Conservation*, pp. 191–202. Berlin: Springer-Verlag.

[55] Tynan, C. T. 1998. "Ecological Importance of the Southern Boundary of the Antarctic Circumpolar Current." *Nature* 392:708–710.

[56] Vaughan, D. G. and C. S. M. Doake. 1996. "Recent Atmospheric Warming and the Retreat of Ice Shelves on the Antarctic Peninsula." *Nature* 379:328–330.

[57] Vaughan, S. 2000. "Can Antarctic Sea Ice Extent Be Determined from Whaling Records?" *Polar Rec.* 36:345–347.

[58] Wilson, P. R., D. G. Ainley, N. Nur, S. S. Jacobs, K. J. Barton, G. Ballard, and J. C. Comiso. 2001. "Adélie Penguin Population Change in the Pacific Sector of Antarctica: Relation to Sea-Ice Extent and the Antarctic Circumpolar Current." *Mar. Ecol. Progr. Ser.* 213:301–309.

[59] Woehler, E. J. 2000. "Recent Trends in Adélie Penguin Populations at Casey, East Antarctica." *Fourth International Penguin Conference, Abstracts*, p. 42. Coquimbo, Chile.

[60] Woehler, E. J., J. Cooper, J. P. Croxall, W. R. Fraser, G. L. Kooyman, G. D. Miller, D. C. Nel, D. L. Patterson, H.-U. Peter, C. A. Ribic, K. Salwicka, W. Z. Trivelpiece, and H. Weimerskirch. 1999. *A Statistical Assessment of the Status and Trends of Antarctic and Subantarctic Seabirds*. Bozeman, MT: SCAR Bird Biology Subcommittee.

[61] Woehler, E. J., D. J. Slip, L. M. Robertson, P. J. Fullagar, and H. R. Burton. 1991. "The Distribution, Abundance and Status of Adélie Penguins *Pygoscelis adeliae* at the Windmill Islands, Wilkes Land, Antarctica." *Mar. Ornithol.* 19:1–18.

[62] Lambert, D. M., P. A. Ritchie, C. D. Millar, B. Holland, A. J. Drummond, and C. Baroni. 2002. Rates of Evolution in Ancient DNA from Adélie Penguins. *Science* 295:2270–2273.

Index

Adélie penguin: age and experience effects on behavior and breeding (summary of), 196–98; breeding synchrony, 107, 144–45, 177, 190, 207, 213–14; central-place foraging, 101 (*see also* foraging); clutch size, 150–53; deferred maturity, 232–38; described and named as species, 5, 7–8; failed breeders, 107, 171; fecundity, 224–26, 230, 238–41; first encounter with humans, 4–7; maturation rate, 129, 136; natural history patterns, summary of, 191–95; navigation, 33–35; philopatry, natal, 89, 137, 221, 268; resilience, 3, 13, 195–96, 198, 207; rookery (*see* colony); subcolony definition, 74, 136; survivorship, 220–24, 230; territory size, 139; wandering, 100, 137; world breeding population, 69, 85–87; yearling, 4, 37, 39, 112–13. See also penguins

— annual cycle, 106–109; response to: food availability, 103–104; photoperiod, 102–104; sea ice, 128; time available, 127–28, 157, 198
— body mass of: adults, 104–105, 125–26, 215; chicks, 172, 212; fledglings, 174
— breeding age, 131–33; incidence, 133–36, 234
— breeding pair: duration, 142–46; formation, 141–42, 157; spread of ages, 142
— breeding success, 151, 181–82, 185–86; effect on: by age, 188–89; by experience, 196–98; by nest location, 187–89; by pair bond, 146, 189–91; by predation, 216–18; by sea ice cover, 96–97
— chick: crèche, 177–81, 214; development, 156, 173, 176; fledging age, 183–84; growth, 173–75; survival, 181–83

— chick care: lack thereof, 182; phases, 176–81; role of sexes, 175
— colony: arrival, 110–11, 128–30; definition, 74, 136, 267; geographic structure among (clustering), 88–90, 221, 250–51, 256–57; growth, 253 ff.; location, 37, 90 ff., 101–102, 127, 260; size range, 87–88, 94; types, 75–85
— colony visitation: duration, 117–19; during winter, 35, 102; factors affecting frequency, 113–19; number of visits, 120–21
— competition for: food, 58–60, 63, 87–88, 89–90, 102, 129, 235, 240, 249; mates, 129, 133, 141–42, 190–91, 234–35; nesting space, 89, 129, 139, 233–34; stones, 139
— diet: during summer (including temporal and geographic variation), 45–51; during winter, 44–45
— diving: aerobic limit and length of dives, 57–58, 62, 65–67; depth, 61–63; effect of light, 35–36, 62, 64; hydrodynamic qualities needed, 55; types of dives, 62
— egg: composition and formation, 155–58; hatching interval, 159; infertility, 169, 216, 224, 237; loss, 166–69; number per clutch, 150–53; size, 150–53
— egg-laying date, effect of: age and experience, 148–50; heritability, 149; interval, 157; synchrony, 147, 149–50; timing, 110–12, 146–48
— feeding proficiency, 129, 153, 235–37, 241, 256; intensity (hyperphagia), 102, 104, 125
— food loads: per dive, 67–68; during summer, 45–48; during winter, 44; necessary to raise chicks, 68
— foraging: pattern, 101; range, 58–61, 89–90; trip duration, 65
— habitat: marine, adaptations to, 63, 256, 268; summer, 37–41; terrestrial (or nesting) requirements, 72–73, 91 ff., 127, 136, 253–54; winter, 25–27, 30–34, 43, 45
— hatching sequence, 172–73; success, 151, 165–69
— incubation period, 158–60; role of sexes, 161–62, 164; routine, 160–65, 168
— migration: autumn, 40–43; premigratory fattening for, 102, 104, 111; spring, 35–37, 97
— molt, 24–25, 30, 32, 40, 43, 105, 112–13, 228; location, 124–25; process, 122, 125–27; reason for, 122; timing, 122–23
— nest: faithfulness to, 137–38, 143–44; location, 137–39, 188; quality, 139–40; stones, 73, 139–40, 253
— nesting period: occupation, 106; reoccupation, 107–108
— population: age structure, 230–32, 242; growth rate, 224
— population size, effect of: food availability, 258; retreating glaciers, 253–54; retreating sea ice, 258
— predation of: adults, 52, 53, 130, 207–209, 235, 237–38, 242; chicks, 177, 179, 182–83, 211–18; eggs, 166–68, 211–18; fledglings, 184–85, 207, 242
— prey capture, 51–55; anatomical adaptations for, 54
— travel efficiency, 129–30; and speed: running, 215; swimming, 56–57, 254; walking, 55, 254
Admiralty Bay. *See* South Sandwich Islands
albatross, 128, 197

Amundsen, Roald, 1–2
Antarctic Circle: definition, 27–29; significance to penguins, 34–35
Antarctic Circumpolar Current, 17; definition, 26; significance to penguins, 28, 30, 35, 256
Antarctic Polar Front, 17, 28, 29, 261; definition, 26
Antarctic Silverfish, 48, 53
Arthur Harbor colony: age, 267; foraging range, 60; layout of penguin nests, 82, 84; leopard seal predation, 207; size, 80, 251–53
Astrolabe, 5–6
auks, 4, 24, 156, 199, 233
Aurora, 35

Balleny Islands, 35, 42, 69, 78
Beaufort Island colony, 93; age, 262; layout of penguin nests, 75–83; size, 78
— foraging range at, 59–60; trip duration, 65
Béchervaise Island colony: appearance of molting birds at, 122; arrival, 111–12; crèches, 177; date of laying, 147; diet, 50–51; foraging range, 58–61; nesting success, 151, 182, 186; occupation, 108–109; wintering area, 30–31, 40; size, 251
booby, 238

Cape Adare colony, 72, 87, 91; area, 82, 84; arrival, 112; chick mass, 172; date of laying, 147; glacial period refuge, 261, 265; layout of penguin nests, 84; size, 78
Cape Barne colony, 93–94, 262
Cape Bird colony, 42, 96; age, 262; chick loss, 183; clutch size, 150–51; crèches, 177; date of laying, 147; diet, 49; egg loss, 167; foraging range, 59; incubation period, 158; incubation routine, 161, 163; layout of penguin nests, 82–83; leopard seal predation, 204–205; nesting and hatching success, 151, 165–66, 181–82, 185–86; occupation, 106–108; pair bonds, 190; size, 78, 250–51
Cape Crozier colony, 2, 12, 89; area, 84; clutch size, 151; crèches, 177; date of laying, 146–47; diet, 45 ff.; egg size, 154; fledgling mass, 174; foraging range, 59–60; incubation routine, 160–61, 163; layout of penguin nests, 82–83, 217; leopard seal predation, 203–209; nesting and hatching success, 151, 165, 181–82, 185–86; occupation, 106–108; pair bonds, 144–45; productivity, 225; size, 78, 87, 221, 230–32, 250–51, 257
Cape Halett colony: arrival, 112; chick mass, 172; clutch size, 151; date of laying, 146–47; diet, 47; egg size, 154; incubation period, 158; incubation routine, 160–61; layout of penguin nests, 217; nesting and hatching success, 151, 166, 182–83, 186
Cape Royds colony, 11, 72, 93, 96; age, 262; area, 137; appearance of molting birds, 122–25; chick loss, 183; clutch size, 151; crèches, 177; egg loss, 167; egg size, 154; fledgling mass, 174; foraging behavior/range, 53, 59–60; incubation period, 158; incubation routine, 161, 163; leopard seal predation, 204; nesting and hatching success, 151, 165–66, 181, 185–86; occupation, 106–109; size, 78, 250–51, 257–58
CCAMLR, 12–13, 43–44, 58, 68, 85
Chappell, Mark, 16

climate change, 13–14, 243 ff., 267–69
colonial breeding, 101–102
continental slope: importance to penguins, 38, 40; relationship to diet, 50–51, 61
Cook, James, 4
cormorant, 197, 201
Crozier, Francis, 2
Culik, Boris, 16

DeMaster, Douglas, 220
Discovery expeditions, 43, 48
Dubouzer, J.-F.-E., 5
Duke of York Island colony, 34, 72, 78, 93
Dumont d'Urville, J.-S.-C., 5
Duroch, J.-A., 6

East Antarctic Ice Sheet (EAIS), 266, 269
East Wind Drift: definition, 26–28; significance to penguins, 33
Emison, William, 45
Emlen, John, 33
Endeavour, 4
Erebus, 2, 19

Falla, Robert, 8
Forster, George, 4
Fulmar, 197, 220

gannet, Atlantic, 128–29, 187–89
glacial ice: effect on nesting habitat, 253–54, 261–69
global warming, 243–44
— response of: ice shelves, 244–47, 253–54; of penguins, 253–58, 262–69; of sea ice, 247–49
gull, 233; black-headed, 187; red-billed, 142, 197; western, 197

Hamner, William and Peggy, 51–52
Hombron, Jacques, 5, 13

Hope Bay colony, 10, 87, 100; arrival, 112; foraging range, 60; size, 80
Hukuro Cove colony, 93, 266; date of laying, 147; diet, 50; egg loss, 167; foraging behavior, 50; nesting success, 151, 181; size, 76, 251–52, 257

ice shelf, 21, 244–47, 253–54, 262–65, 269; definition, 93
Inexpressible Island. *See* Terra Nova Bay colonies
International Geophysical Year (IGY), 11

Jacquinot, Honoré, 5, 13
John Biscoe, 10

King George Island. *See* South Sandwich Islands
kittiwake, black-legged, 128–29, 142, 187–89, 197–98, 220, 224, 234, 238
Kooyman, Gerald, 16, 32
krill: Antarctic, 43 ff., 52, 203, 248, 258; crystal, 48 ff.; predator avoidance, 51–53

Last Glacial Maximum (LGM), 90–92, 243, 249, 260–67
LeResche, Robert, 12
Levick, Murray, 34, 69–72, 99, 136, 140, 178, 185, 195–96, 202, 203
Linnaeus, 4
Little Ice Age, 267
Lützow-Holm Bay. *See* Hukuro Cove colony

MacRobertson Land. *See* Béchervaise Island colony
Magnetic Island colony, 64; diet, 49–50; foraging behavior, 53; leopard seal predation, 203–209; nesting success, 151, 186

Matthews, L. Harrison, 9–10
Mawson, Douglas, 53–54
Mount Biscoe colony, 91–92
Murphy, R. C., 8
Murray, James, 9, 178, 181
Myctophid fish, 45, 47

Neider, Charles, 2

Ogilvie-Grant, William, 7

penguins (other than Adélie): African, 4–5, 16, 45, 201, 202, 220, 233, 239–40; chinstrap, 11–12, 29–30, 31–32, 58, 63–64, 91–92, 113, 127, 130, 136, 155, 159, 191–96, 257, 258; crested, 16, 29, 103, 128, 156, 187, 201, 220, 235, 241, 257; emperor, 10, 11, 16, 19, 29, 32, 63, 73, 96, 97, 103, 201, 203, 241, 249, 257–58; Galápagos, 104–105; gentoo, 7–8, 11, 58, 63, 91–92, 103, 113, 127, 128, 130, 136, 155, 159, 169, 191–96, 259; king, 11, 29, 201, 220, 241; little blue (or blue), 29, 156, 197–98, 220, 239–40; Magellanic, 5; yellow-eyed, 12–13, 29, 128, 142, 148–49, 153, 169–70, 197, 220, 224, 234, 239–41
Pennell, Harry, 200
Penney, Richard, 33
peregrine falcon, 199
petrel: Antarctic, 16, 37; diving, 199; snow, 16, 37; storm, 199
Point Géologie colony, 5–7; appearance of molting birds, 122; occupation, 108; size, 251
polynya: definition, 35–37, 95, 97
— effect on: colony distribution, 94 ff., 265; incubation routine, 160
Possession Island, 7, 87

Prydz Bay. *See* Magnetic Island colony
Pucheran, Jacques, 8

Reilly, Pauline, 74
Richdale, Lance, 12
Roberts, Brian, 8
Ross, James Clark, 2, 7, 19, 21–22, 56

Sapin-Jaloustre, Jean, 11
SCAR, 74, 81, 85
Schackleton, Ernest, 9, 11; Keith, 13–14
Scott, Robert, 10
sea ice: affected by climate change, 13–14, 247–49; annual cycle of, 23–25; as impediment to travel, 21–24, 35; as penguin habitat, 16–18, 22, 25 ff., 254–58; refugia, 24–25, 26, 257; types, 18 ff.
— extent: minimal, 27; maximal, 17, historic, 248–49, 261
sea ice variation, effect on: annual survival, 242, 256; clutch size, 150; colony arrival, 115–17, 240; colony location, 23–25, 34, 43, 92–94, 253–54; colony size, 221; egg laying, 147; egg size, 153, 155; incubation, 160, 163; nesting success, 165–68, 185–86; penguin natural history patterns, 191–96; seal predation, 205–207, 257; skua predation, 217
seal: crabeater, 32, 202; elephant, 201, 205; leopard, 3, 19, 52, 56–57, 130, 203–207; Weddell, 204
shearwater, 239; Manx, 128–29, 149; short-tailed, 128, 149, 197, 220
Signy Island. *See* South Sandwich Islands
skua: brown, 11, 210; south polar, 16, 128, 142, 166, 177, 178 ff., 197, 210–11
— factors: that encourage, 211–13; that constrain, 213–16

Sladen, William, v, 8; contributions of, 10–12, 99–100, 269
South Orkney Islands colonies, 4, 5, 10, 39–40, 44, 191; appearance of molting birds, 124; arrival, 100, 111–12; clutch size, 150–51; crèches, 177; date of laying, 146–47; diet, 50; egg laying, 100; size, 154–55; fledgling mass, 174; foraging trip duration, 65; incubation period, 158; nesting and hatching success, 151, 166, 181–82, 186; occupation, 108–109; pair bonds, 144–45; relationship to sea ice, 25; routine, 160–61, 163; size, 251–52
South Sandwich Islands, 261
South Shetland colonies, 4, 5, 31, 191, 266; clutch size, 151; diet, 50; fledgling mass, 174; incubation routine, 160–61, 163; nesting and hatching success, 151, 165, 181–82, 186
— foraging behavior/range, 58; trip duration, 65
Stonehouse, Bernard, 10–11
study methods: marine ecology, 16 ff., 45, 58, 62–63, 66; marking, 10–12, 226–30
Syowa Coast. *See* Hukuro Cove colony

Taylor, Rowland, 11
tern: Arctic, 197; sooty, 128–29
Terra Nova, 200
Terra Nova Bay colonies, 72, 95–96, 185, 204, 207, 262–65, 267
Terror, 2

Vestfold Hills colony, 87

West Antarctic Ice Sheet (WAIS), 261–69
whale: baleen, 28, 43–44, 52, 248–49, 258; killer, 200–203; minke, 19, 28, 32–33, 249
whaling, 43
Wilson, Rory, 16, 45
Windmill Islands colonies: age, 266; appearance of molting birds, 123–25; layout of nesting, 85–86; occupation, 108; pair bonds, 144, 190–91; size, 251
winter water, as Adélie habitat, 26, 40
Wood, Robert, 12

Young, Euan, 211–15, 218

Zélée, 5